▶▶▶ Rave Reviews for The Golden Corridc

"... lots of authentic, historical pictures. ... a wonderful job of putting this book together.

... you will actually feel the joy, anguish and determination of the brave people who settled this land ...

"*The Golden Corridor*" is beautifully organized ...

... **this is the one book you need to have.** You will spend hours reading the writings of those who were here and delight in the restored pictures of that era."

Mountain Democrat

"Fans of old photographs ... will love "The Golden Corridor." The book is filled with amazing black and white photos that bring early Northern California to life.

Sidebars on each page give fascinating quotes from diaries, journals and newspapers, as well as anecdotes.

I am a lifelong resident of Northern California, and I learned from this book ... "The Golden Corridor" is well worth checking out."

The Union, Nevada County

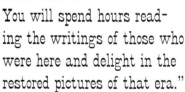

...this is the one book you need to have. You will spend hours reading the writings of those who were here and delight in the restored pictures of that era."

Mountain Democrat

"... a captivating study of 19th century people who helped shape the times.

Sacramento Bee

"... this amazing collection of firsthand testimony ... Sidebars offer amusing quick vignettes from the era! **Enthusiastically recommended reading. . ."**

Midwest Book Reviews

"... **educational and entertaining.** Profusely illustrated ..."

Auburn Journal

THE GOLDEN QUEST
And Nevada's Silver Heritage

Includes the Lake Tahoe Region:
Truckee, Tahoe City, Reno, Carson City, Virginia City, Genoa, Glenbrook
and smaller California and Nevada communities within 30 miles of Lake Tahoe

Written by
California and Nevada pioneers and 19th century historians.

Researched and compiled by Jody and Ric Hornor
Photo restoration by Ric Hornor and Steve Crandell
Edited by Paula Bowden

Copyright © 2006

Ric and Jody Hornor

ISBN # 0-9766976-4-5

First printing: May 2006

Published by

Century Books
An Imprint of Electric Canvas™
1001 Art Road
Pilot Hill, CA 95664
916.933.4490

www.19thCentury.us

Acknowledgments

This book is dedicated to the thousands of pioneers who wrote California and Nevada's history, the early photographers who captured the scenes, and the hundreds of historians who have preserved it over the years.

This work would not have been possible without the help of dozens of libraries, librarians and archivists including:

Special Collections, University of Nevada, Reno

Nevada Historical Society

Northwestern University

The Library of Congress

Denver Public Library

Bancroft Library

Steve Crandell

As we searched the vast libraries for the appropriate photographs, we found many of the same photos in different archives. The attributions are based on the archive from which we actually obtained the image. Our apologies for any real or perceived errors.

Map of the Lake Tahoe Region, 1895

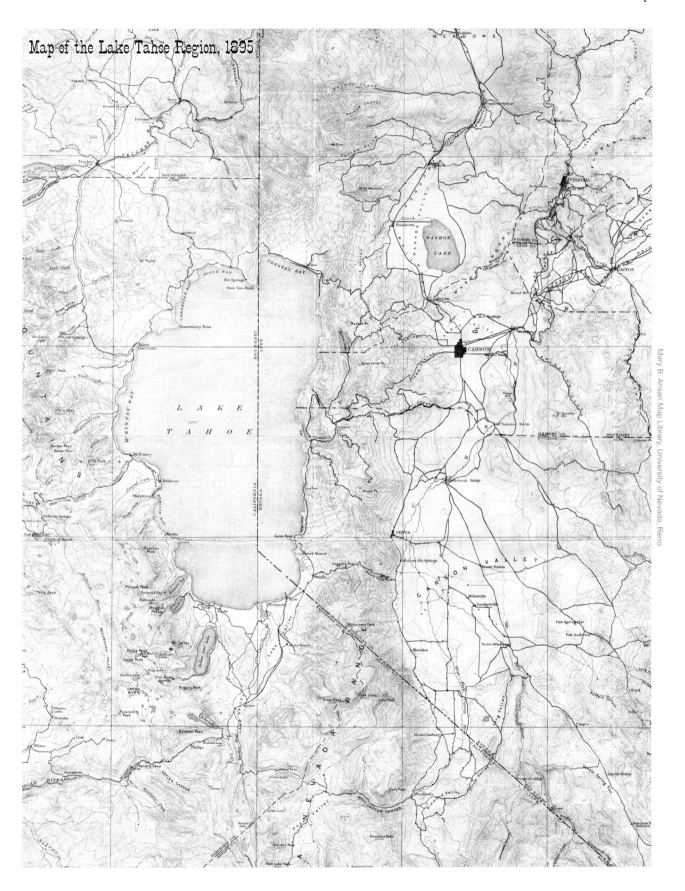

This log cabin, located in Genoa, was the oldest building in Nevada, and known originally as the Mormon Station. Photo 1868.

Introduction

Trappers, adventurers, pathfinders, emigrants. . . even before the Gold Rush began there was a trickle of the truly hearty, wildly adventurous and amazingly tenacious people who came and began to settle Nevada and California. Seeking a new life, they braved the elements, wildlife and dangerous terrain to stumble upon one of our nation's greatest natural treasures, Lake Tahoe.

But with the discovery of gold in 1848 and the subsequent Gold Rush in 1849, Lake Tahoe's beauty was enjoyed for but a moment in time as they "rushed" by it to the gold mines of California.

After ten years in California, the next "quest" was for gold in Nevada. But as we know, all that glitters is not gold. The quest quickly turned to silver and Nevada's rich deposits of the dazzling mineral.

The new "rush" to Nevada's mines began just ten years after California's Gold Rush. Many of the miners once again

"... there was a trickle of the truly hearty, wildly adventurous and amazingly tenacious people who came and began to settle Nevada and California . . .

passed through the Lake Tahoe area, and it once again became a brief resting place for the weary travelers.

The rush to Nevada's mines created even more demand for lumber . . . for homes, businesses, and timbers to shore the mines.

Hillsides were stripped of trees. Flumes were built to transport logs. Lumberjacks with their ox or mule teams worked day and night. Lumber mills popped up everywhere and were months behind in their ability to supply the voracious needs of the rapidly growing California and Nevada communities.

It wasn't until the 1880 that people truly became awed by Lake Tahoe's beauty and enraptured by its treasure of resources that it enticed large numbers of people to settle around the lake, create resorts, and tap into its beauty and vitality for their own enjoyment.

In *The Golden Quest*, you'll read the area's most enlightening, entertaining and significant history culled from roughly 3000 pages of detailed source information. It gives you a glimpse of the many monumental accomplishments, brave souls, and exciting times that made the Lake Tahoe and Western Nevada region what it is today.

As you read, keep in mind that with the exceptions of the photo captions, introduction and preface, this book was written in the 19th century. You'll find the key points of history delivered in the colorful language of the time as each individual contributor's writing style has its own character.

Also, as you read from chapter to chapter, you may find different formats and writing styles. That's because the main body of text in each chapter may be from its own unique source.

The diaries, journals and other works quoted in the sidebars of this book were written very differently as well. Words were spelled differently in those days, and punctuation and sentence structure were often different as well. Some are challenging to read but well worth your effort. Our goal is to preserve these styles for you to enjoy, so we have intentionally done nothing that will make the chapters consistent in their presentation, punctuation, spelling or format as it would eliminate each contributor's personal style and flavor in doing so.

...key points of history are delivered in the colorful language of the time as each individual contributor's writing style has its own character.

There were huge cultural biases that are reflected in the text. As ingenious as these brave pioneers were, they had yet to invent "political correctness." As degrading and disheartening as some of the terms and stories are, they do reflect history.

Keep in mind that history, as it was recorded in the 19th century, was often done so subjectively. Many of the county history books (from which much of this text is taken) were paid for by the support of the people whose lives were chronicled within them. Thus the poor, or not so vain, may have been omitted, unless they were truly newsworthy.

We're fortunate that there were a number of photographers, especially after 1860, that traveled extensively through the area and took hundreds, if not thousands, of photographs. Even so, finding the one image that exactly illustrates a point in the text is often impossible. We must sometimes go out of the specific geography or era to give you the best illustrations of the places and events being discussed.

Many of the original photographs are so damaged that it's nearly impossible to see the detail in them. They are carefully restored to uncover details that have faded over the years. In some the detail we uncover simply amazes us, like the American flag that is attached to the Indian headdress on page 119. In others there's a tinge of disappointment that we can't fix the years of degradation. Even so, some of those images are included because they illustrate important points of our history.

Hundreds of hours of restoration have gone into the photographs used in *The Golden Quest*. Many of the original photos are in the public domain and available through the organizations noted with each image. Some are from private collections. In all cases, the restoration work is copyrighted. If you'd like copies of these images for your own use, you must contact the libraries or archives noted and obtain them directly from the respective institution.

So have fun wandering through the Lake Tahoe and Western Nevada regions in the 19th century. Get a taste of the rich history. And, if this book has whet your appetite for more, pick up a copy of *The Golden Corridor*, which covers the area from San Francisco to Lake Tahoe. Or use the bibliography as a guide and find yourself a library with a good California and Nevada reference section. Have fun with your studies!

Sidebar Legend

Most of the side bars provide new information that is not contained in the main body of text. The nature of the information is denoted by the following icons:

Text from personal letters will appear in this script font with the quill pen.

Stories about crime will appear with the hangman's noose in this font.

Quotes from diaries or journals will appear with this image of President Taft's personal journal.

Call outs, text that appears in or comes from the main body of content, are highlighted for emphasis and appear with this magnifying glass.

Old newspaper articles will look like this and include the newspaper name and date, if known.

Table of Contents

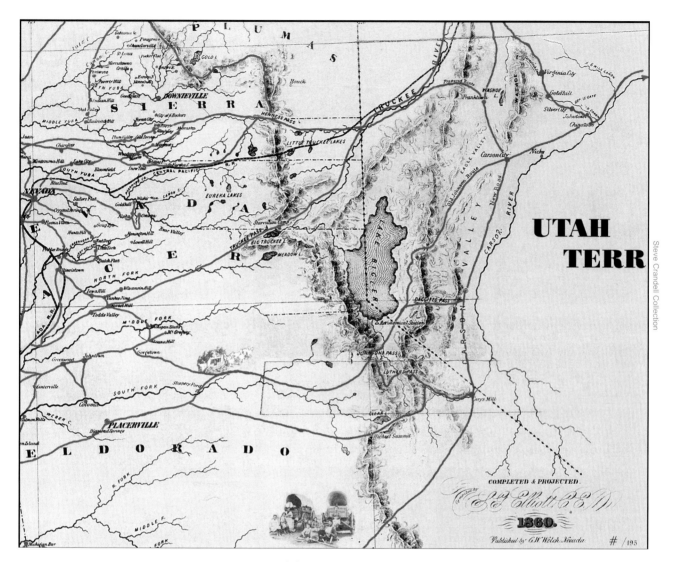

Map of the Utah Territory, 1860.

Tallac Hotel and guests. Although the photo caption for this image claims the date as "1886," E.J. "Lucky" Baldwin built the Tallac Hotel in 1899. It was very modern with indoor plumbing and electricity. It had an intricate system for piped water from Fallen Leaf Lake. The electric lighting "turned night into day." Fountains arched from pools throughout the gardens and manicured flower beds bordered gravel walkways.

Chapter 1: El Dorado County – Lake Tahoe, the High Sierra and vicinity

LAKE TAHOE is located on the eastern side of the central ridge of the Sierra Nevada. According to the observations of the United States geographical surveying corps, under command of Lieut. George M. Wheeler, the altitude of the lake is 6,202 feet above the level of the sea; that of Tahoe City, 6,251, and of Hot Springs, 6,237 feet. The water of the lake being shed from the solid granite and volcanic mountains that compose its boundaries by more than thirty streams, is extremely pure and clear, and when in a state of quietness, one can observe fish and other objects most distinct and perfect to the depth of from thirty to forty feet; it is of blue color and very cold, but never freezes in the winter.

Our attention will be drawn next to the timber supply of the great forests. In this respect the county certainly is not behind any part of the State, if we except the redwood forests

"An opening in the trees, a turn in the road, and Lake Tahoe is before me. Not a ripple on its surface. Surrounded on every side by snow-clad hills, whose sides are covered with pine forests, all of which are reflected as in a mirror, it looks like a painted lake. There is a sense of mystery in its unfathomable depths, a feeling of awe at this volume of water suspended six thousand feet in the air, never varying in its height, never frozen over like neighbouring lakes."

J.G. Player-Frowd, 1872

I had quite a notion I would like to go mining again, and when the excitement broke out over the mountains at Washa, Virginia City and Gold Hill, I went over there the year after the first excitement broke out, and that next winter of 1861-1862. My mining venture did not amount to much. I returned to California and had rather a severe time crossing the mountains, as the snow on the summit was over twenty feet deep. The sleighs ran part of the way, but could not cross the summit, but three of us were footing it. In going from Lake Tahoe to the next station, soon after passing Lake Tahoe Hotel, in crossing a small lake or lagoon, we broke through the ice, and the water took us up to the waist. When we got out, two of us stopped to wring the water out of our clothes, but the third poled right ahead and left me with the other man who soon gave out worrying through the snow, and I could not get him along. He would want to rest about every fifty yards. Finally night came on, and by this time I had to almost carry the man by letting him lean his weight on me by placing his arms over my shoulders. It was not long until I knew that I could not stand the burden much longer, and finally he lay down in the snow, and I knew I could not get him up or even try to

continued in far right column

A little farther north near Blue Canyon, hotel guests had to tunnel their way in and out like gophers as well.

of the Coast Range, which monopolize with their product the market of San Francisco. The demands of the miners have practically divested the western half of the county of the timber for the manufacture of lumber, but there is no limit to the supply for fuel anywhere, while the new growth will soon cover the vacant lands with all the timber required for any purpose. Further east, excepting the highest peaks of the Sierra Nevada, the country is covered, with a dense growth of the finest timber in the world. We believe we are safe in saying that El Dorado county has, to-day, not less than 600 square miles of virgin forests. This consists principally of cedar, spruce, fir, several varieties of yellow pine, and the magnificent sugar pine. In the higher altitudes, tamarack is found in large quantities, while an occasional hemlock puts in an appearance.

Along the shores of Lake Bigler, and far back toward the mountain tops, the timber is being rapidly cleared away, to supply the Virginia mines and the Nevada towns in general. What the annual cut in that region is, we are unable to state. It is run into the lake and towed in rafts by steamers to Glenbrook, whence a narrow-gauge railroad has been built to carry it over the mountains.

INTERNAL IMPROVEMENTS.

ROADS. The old emigrant road entering the State and County by the way of Carson valley; the old Mormon station was considered to be the first trading post this side of the State line; from here the road crossed the summits of the mountains, then turning around the southern end of Silver Lake, it descended passing between the head waters of the American and Cosumnes rivers, following the divide between these rivers through Sly Park, Pleasant valley, to Diamond Springs, and from there to the low-lands by the way of Mud and Shingle Springs, Clarksville and White Rock Springs into Sacramento county. This old emigrant road, or rather the "emigrant route," traced and recommended in all the guide books, and by the-footprints of annual migrations to the State, for eighteen years, passed through El Dorado county from east to west, her entire length, branching off from Grizzly Flat south to Brownsville, Indian Diggings and Fiddletown; from Diamond Springs via Placerville to Coloma, Kelsey's, Spanish Flat, Georgetown, Greenwood, Centreville, Salmon Falls and all points of the northern part of the county; from Mud Springs to Logtown, Saratoga and Dry town; from Clarksville to Folsom.

Hunt, a Mormon sent out from Salt Lake in the spring of 1849, as an advance agent for the Mormons, to explore the Sierra Nevada for a route to be traveled with wagons, started out with fifteen or sixteen men and several wagons and selected the route, which, with slight modifications, was traveled after him by thousands and thousands of immigrants; a very large proportion of their number, for the period of eighteen years,

North American Hotel on the western summit, Placerville Route, 1860s

continued from far left column

get him to help himself any more. By this time we had just gotten in sight of the lights in a house. He was down in the snow and I was rubbing him with all my might, and if I left him to go for help, I knew he would be dead before I could get back with help. Just at that time I saw a dark object in the road approaching and it proved to be the mail stage, and we dumped him in and I told the driver to go just as fast as he could, so we soon got him to the hotel, but he was entirely unconscious and had been for some time. Some time after getting him to the fire and a lot of stimulants taken inwardly with some smart rubbing on the outside, we brought him to.

When arriving at the summit we could only see smoke coming out of the snow here and there. Even the chimneys did not reach through the snow. People were living down there under the snow and had run tunnels from house to house like a lot of gophers.

Lorenzo Dow Stephens, 1849.

Yank's Station was the Pony Express remount station 143 in what is now known as Meyers, California. It was operated until October 26, 1861. George D. H. Meyers bought it in 1873.

Our road led us around and partially over a very rough mountain and down again into a ravine, (a tributary of the lake), and then we commenced the long ascent of the mountain the top of which is the highest summit we have to pass, being 9000 ft above the sea. We all walk in going up these steep hills and sometimes in going down. The other seven of the party joined us in the summit but their we again divided size keeping the main road which here makes a wide circuit, and four taking a shorter but rougher trail. At the summit we passed between great snow drifts but the air was mild and clear. Extensive fires were raging along the western slope of the mountains and the atmosphere was so filled with smoke that our great view of Cal. or at least the Sacramento valley from the summit was entirely spoiled.

William Henry Hart, September 17, 1852

first interrupted their westward journey to try their fickle, varying fortunes here within the limits of the Empire county, where the gold was discovered that had caused the immigration to this El Dorado.

To give the reader an idea of the travel over these roads in early days, we may quote from the register of immigration, kept by Mr. J. B. Ellis, the aggregate of wagons and animals that passed over the mountains into California, by the old Carson route, during the summer season of 1854, commencing on the first of July that year, amounted to: 808 wagons, 30,015 head of cattle, 1,903 horses and mules, 8,550 sheep.

The immense proportions to which this immigration was grown, caused others to find another and possibly easier accessible route to compete with those in existence. And a third claimant for a newly explored mountain route across the Sierra Nevada from Carson valley into California and particularly El Dorado, turned up in the person of a Mr. Dritt, by mountain men of that time generally known as 'Old Daddy Dritt.' A petition was presented to the State Legislature in session, in April 1854, for a charter for a wagon road to connect the Carson valley with Placerville. This petition was signed by Messrs. Dritt, Murdock & Co. This new route by which the steepest ascent would be avoided and which therefore presented an eligible road for wagons, was designed to commence at the mouth of Big canyon on the old Carson river route, intersecting the Johnson Cutt-off road on the western summit of the mountains and to cross the South Fork of the American river at Bartlett's bridge. Mr. Dritt was an old experienced mountaineer, who had crossed the Sierra Nevada frequently, and himself as well as other people had great confidence in this route as the easiest pass that could be found.

The prospect of still more competition to that one going on already by the Johnson Cutt-off road, and the advantages that were offered to the traveling people by using these roads, stirred up the settlers of Carson valley and further along the Emigrant road, and made them afraid to loose the trade.

NATIONAL WAGON ROAD. A Yankee was proposing to establish a regular mail line between California and the Eastern States via St Louis. Mr. Wm. N. Walton, in April, 1855, presented to the State Senate of California a memorial in which he makes the proposition that the State Legislature of California should by legislative act donate to him (Walton) right, title and interest of the State in and to certain quarter sections of land (not to exceed five quarter sections) situated between the eastern boundary of the State and the Pacific coast, as stations for the encouragement of an overland immigration by means of camels or dromedaries.

The above mentioned Board of Commisioners, after a thorough examination of the different routes in September

Logging at Lake Tahoe

Steve Crandell Collection

June, 1864 — Sheriff Rogers was informed of the [stage] robbery, and he and several attachees of the stage company, started in pursuit of the robbers. Sheriff Rogers, with Taylor and Watson, arrested two men at the Thirteen Mile House. They had taken supper the night before at the Mountain Ranch, but left and called between 12 and 1 o'clock in the morning at the Thirteen Mile House, asking the proprietor to allow them to sleep in his stable. On his answer, that he did not allow anyone to sleep in his stable but he told them they might sleep up stairs in his house, and they accepted the proposition. For concealing their countenances they had drawn their hats over their faces while talking and entering the house. In the morning they overslept themselves and were arrested while in bed, brought to Placerville and lodged in jail.

Meanwhile deputy Sheriff Staples and Constables tracked the robbers to the head of Pleasant valley. Staples inquired of the landlady if there were any men in the house, and she replied; "Yes, six, up stairs." He rushed up stairs, seized a gun standing at the door of a sleeping room, burst the door open and presenting the gun cried: "You are my prisoners!" But scarcely had he uttered these words, when the robbers fired, wounding him fatally, he fired at the same time, hitting one of the robbers in the face. Officer Ranney, also, was dangerously wounded, both officers were robbed by taking their money, watches, horses and arms; whereupon they decamped, leaving their wounded companion behind.

ROBBERY AND MURDER AT PERU

On the evening of October 20th, 1860, while four miners of the vicinity were seated in the store of Messrs. Pierson & Hackamoller, engaged in a social game of cards, five men with masked faces and pistols in hand entered the store. The first party, supposing that they were a party of miners, bent on a little fun, attempted to set the dog on them, which move was responded by the robbers with a shot, fired at the card players, and the advice if they would remain quiet, they should not be hurt. Upon this proposition being agreed to, they demanded of Mr. Pierson the key to his safe. He told them it was not in the store; whereupon they commenced to beat him with the butt end of their pistols, he warded off the blows and tried to make his escape by a door leading into the family room, which they were determined not to allow him. He was fired upon by one of the villains, the shot entered near the eye, producing almost instant death. Then they took the key from his pocket, and rifled the safe of its contents, and departed. The safe at the time contained a thousand dollars or more. This robbery and murder, unequalled for boldness and daring, produced great excitement, Mr. Pierson being one of the best and most respected citizens.

Teams climbing the grade at Slippery Ford, 1860s

1855, reported in favor of the route along the South Fork of the American river, passing Slippery Ford, Johnson's pass, Lake Bigler, Luther's pass, Hope valley and Carson canyon to Carson valley. Under date of October 16th, of the same year, the Board of Commissioners advertised for sealed proposals for the construction of a wagon road over the Sierra Nevada by way of the above named places, according to plans and specifications, etc., but nothing was done against active work on the road.

BRIDGES. The many streams of perennial running water, having their sources high up in the Sierra Nevada mountains, as we have seen before, for a few months in the year only enable a fording at one or another spot, while for the greater part of the year the high stand and the rapid flow of their waters necessitate some other means to carry the travel across. The pioneer emigrant road of El Dorado county winding itself down from the mountains, following the divide between the Middle Fork of the American river and the head waters of the North Fork of the Cosumnes, piloted through by the Mormon Hunt, is the only road that avoids all the larger streams and enables a trip from Silver Lake down to Sacramento without crossing one stream of water that amounts to anything. Traveling on all the branch and cross-roads, leading off the former on both sides, however causes traversing one or another of the larger or smaller rivers that roll their waves down through this county finally to empty into the Sacramento river.

As the first device, to assist the traveling people on said roads across the natural waterways, ferries of the most primitive make up and clumsiest construction and shape were in use; old ship's boats of all sizes had been pressed into the service or an ingenious fellow had accomplished the same purpose by transforming some old emigrant wagon-beds that had come all the way across the continent,

Sixteen and five-eighths miles east of Placerville, where the Johnson's Cut-off road crossed the South Fork of the American river, was Bartlett's bridge, carrying a great part of the emigrant travel across the rapid stream. It was a heavy wooden structure, but could not resist the force of the high water which came down in torrents on March 7th, 1855, and was swept away. The communication thus interrupted for a while, caused the travel to go the other route. Then B. Brockless took

Webster's Station, Sugar Loaf Mountain. Webster's Station appears on the 1861 mail contract as Pony Express station 145. The station, which stood on the Placerville-Carson Road, began as an original C.O.C. & P.P. (Central Overland California & Pike's Peak) Express Co. station in April 1860. It also served as a stop for teamsters and the stage lines until the late 1860s. Travelers also knew Webster's as Sugar Loaf House from a nearby rocky mountain of the same name. Photo 1860s

Camped to night near a trading post called the Mountain House where we bought Hay at 10 cts per lb., Barley at 15 cts Bread at 40 cts + Beef at 25. Their appears to be scarcely any grass in these mountains and cattle must subsist chiefly on the browse or the tender twigs and leaves of trees.

William Henry Hart, September 18, 1852

Courtesy of the Library of Congress, Lawrence & Houseworth Collection

The Carson Tahoe Lumber & Fluming Company, Glenbrook, Nevada, 1885-1890.

HIGHWAY ROBBERS ARREST.

Three desperate fellows, giving their names as Faust, De Tell and Sinclair, started from Sacramento in the later days of July, 1867, with a determination to make money some way. They commenced by robbing houses along the road, and on Tuesday, August 3d, stopped a teamster on his return from Carson Valley, just above Sportsman's Hall, and made him shell out; then coming up the road, robbing houses at their pleasure, also picking up a man who was driving a water cart on the road, for ten or twelve dollars. Under-Sheriff Hume, with a posse of three or four men, went in their pursuit, and being informed of their course between the time, by Constable Watson, of Strawberry, he lay in wait for them at a point in the road near Osgood's toll house which they could not well get around. About half-past eleven on August 5th, the robbers came up all armed with rifles. Hume ordered them to stop, whereupon one of them fired, the shot taking effect in the fleshy part of Hume's arm,

continued in far right column

up the idea given by Sherman Day, who some time previous, surveying on the State road line, had designated a point, a few miles further up, as the place where the road ought to cross the river. Here a bridge was soon built, known as Brockless bridge.

STAGE LINES

In June, 1857, when the first work for improving the Johnson's Cut-off road, across the Sierra Nevada from Placerville to Carson valley, was just commencing, the Board of wagon road directors made an inspecting trip over the said road, on which occasion the pioneer stage-man of the Pacific slope, Col. J. B. Crandall, took one of his six-horse Concord stages over the mountains, with the intention to start a weekly stage between Placerville and Genoa, which was altered to a semi-weekly stage line on May 18, 1858, running as an overland mail line from Placerville to Genoa, Carson valley, Sink of the Humboldt and Salt Lake City. The passenger fare from Placerville to Salt Lake City amounted to $125.00. This, however, was only the embryo of the great

OVERLAND MAIL LINE. About the middle of June, 1857—Mr. Theo. F. Tracy opened a tri-weekly express line from Placerville to Genoa in Carson valley, connecting at Placerville with Wells, Fargo & Co.'s express and running with Crandall's stage line, just then started via Sportsman's Hall, Brockless' bridge, Silver creek, Lake valley, Hope valley, Gary's mill and Mormon station.

SNOW SHOE THOMPSON.

John A. Thompson, better known as "Snow-shoe Thompson." He was a Norwegian by birth, and the first to introduce a Norwegian pattern of snow-shoe. A pair of them can be seen at the present time at the Ormsby House, in Carson City.

Hearing of the difficulties attending the transportation of mail across the Sierra on account of the great depth of snow, he determined one day to make a pair of snow-shoes such as he remembered to have seen when a boy in Norway. Having made the shoes, he went to Placerville, near which place he could practice using them and test their utility. Finding that they worked to his entire satisfaction, he undertook to carry the mail across the Sierra on them, making his first trip in January, 1856. The distance, ninety miles from Placerville to Carson Valley, was passed over in three days, the return taking one less because of the down grade. Having made the experimental journey successfully, Thompson continued to carry the mail between the two points all that winter. The weight of the mail bags was often from sixty to eighty pounds. When traveling across the mountains he never carried blankets or wore an overcoat. He traveled by night as well as by day when necessary. If he camped for the night, he hunted the stump of a dead

Teamsters at Slippery Ford in the 1860s, ready to deliver their loads to their final destination.

Courtesy of the Library of Congress, Lawrence & Housewoth Collection

continued from far left column

though not hurting him seriously. Hume then ordered his men to fire, and when the smoke cleared away they found two of them lying on the ground, one being dead, the other unhurt; the third one had been seen falling off the bridge, and until the next morning was believed to be drowned in the creek; but then they found that he had recovered and crawled under the bridge, where he stayed until all were in the toll house, when he—minus two coats—started back towards Placerville. One hour after daylight the Sheriff's party struck his track, and he was captured a short distance above Brockless' bridge, and both the prisoners brought to Placerville and lodged in jail. Before Court Sinclair stated: My name is Walter Sinclair; am one of three men that were in the party that fired upon the Sheriff's party; am from Arizona; served there under Gen. Conner; am from New York; aged 21 years. The dead man was a German by the name of Faust; age unknown; was deceased and another man named Hugh De Tell. Their trial ended with a sentence for a good long term to be sent to the State prison.

While we do not wish to depreciate the services and merits due to Thompson, it is due to truth and justice also to state, that one of the earliest settlers of El Dorado county, Jack. C. Johnson, of Johnson's ranch, preceded Thompson as a trans-mountain mail carrier; he was the man that opened up, marked out, and traversed the route called after him, "Johnson's Cut-off," which subsequently was traveled by Thompson. By this very route, and through this pass Johnson has carried the mail from the present site of Genoa to Placerville in twenty-six and one half hours previous to Thompson's first trip over the same route. It is not more than right that the government appreciated Thompson's services who intrepid and faithfully did his difficult and dangerous duty, unconcerned of season and weather, but let the truth of history be vindicated. Jack Johnson claims the name as the Nestor and pioneer of trans-mountain mail carrying on foot by the Placerville route.

Courtesy of the Library of Congress, Lawrence & Houseworth Collection

In 1856, a Mr. Swift and Mr. Watson began operating the Strawberry Valley House as a hostelry. In 1859, a Mr. Berry managed station operations at Strawberry, and he established a partnership with Mr. Swan to build a road over the mountain. Berry also served as the station keeper when the Pony Express began. One source suggests the station's name, Strawberry, came from Berry's alleged practice of feeding travelers' horses with straw, while the owners had paid for hay.

pine tree and having set fire to it, he built him a bed of spruce boughs, on the snow, and lying down with his feet to the fire rested and slept soundly. He was never lost in the woods or the mountains. By observing the appearance of the trees and rocks he could tell which way was north and which south and direct his course accordingly. He helped to bring the material over the Sierra Nevada mountains on which the Enterprise was first printed at Genoa in 1858. He was in the battle with the Pah-Utes in May, 1860, at Pyramid Lake, when the whites were routed with great slaughter.

He was a man of great physical strength and endurance, and of such fortitude of mind and spirit, that he courted, rather than feared, the perils of the mountains when visited by their fiercest storms; and the wild rage of a midnight tempest could not disconcert or drive him from his path. But under the strain of the exhausting labors be forced upon himself, his great strength gave out, and in the prime of life he was compelled to surrender to Nature's last summons. After a brief illness, at his residence in Diamond Valley, he died May 15, 1876.

Genoa in the 1870s. The Raycroft Hotel is far left and their stables the next building down. You can also see the Gilman Hotel, Odd Fellows building and the court house.

Chapter 2: Douglas County - Genoa and vicinity

Genoa, the county seat of Douglas County, is among the oldest settlements of Nevada. The locality first attracted the attention of some Mormons in 1848, who were en route to the gold diggings in California from Salt Lake City, and during that and the succeeding years a few families settled here. Not until 1850, however, did it assume the title of Mormon Station, by which it was so well and widely known for many years afterwards. As a trading and recruiting place for the immense emigration which was then flocking toward California from the East, its situation was admirable, while its pastoral advantages were great, and the adaptability of the soil to the culture of grain and the hardy vegetables had been proven.

THE FIRST TRADERS. Principal among the Mormon settlers of Genoa was Col. John Reese, after whom Reese River was subsequently called, a man of robust energy and much enterprise. He started the first trading-post, and also fed the hungry emigrants for a consideration. But he did not stop at these. He put up a blacksmith shop and shod their animals and repaired their wagons, and later erected a flouring and saw-mill.

About eighteen miles south-east of Genoa, in the lower hills of the Pine Nut Mountains, in the fall of 1859, there were found such mineral "indications" as to create considerable excitement among the residents, and many claims were located. The following year, however, more encouraging "prospects" were obtained higher up in the same range, which caused an abandonment of the first-named discoveries, and the immediate formation of Eagle Mining District.

Raycroft Hotel, Genoa, Nevada, 1882

 Our road yesterday was pretty much the same. While staying here, a waggon drawn with 4 horses, and with others to relieve them, passed by us. It belonged to a merchant that had in the summer taken some goods to be disposed of in Utah, whom managed, as he told some of the brethren, to make a pretty good business of it. He carried passengers also, of which he had 3, now returning from the Washoe Diggings, having made very good time of it. One of them had a nugget upwards of 1 lb. weight, and numerous smaller ones.

William Ajax, 1861

THE PIONEER HOTELS. After Colonel Reese, the first hotel was kept by a man named Merkly, who, after awhile, sold out to George W. Brubaker, and he, in turn, disposed of the establishment to a man named Raycroft. Mr. Brubaker subsequently erected the building in the north end of town known as Rice's Hotel.

EARLY CONDITION OF GENOA. At the time of the discovery of silver, there had congregated in the immediate vicinity of Genoa about 200 people, the most of whom had been attracted there by the agricultural and grazing advantages which the locality possessed. Several hundred people were engaged in gold mining to the southward at what was then called the Walker River, or Mono, mines, who, to a considerable extent, made Genoa their supplying point. Genoa also had a newspaper; it was connected with the outside world by a telegraph line, and the overland stages passed through. A grist and saw-mill were in operation. Two stores supplied the residents with all necessaries in the grocery and clothing lines, and it was the leading town of western Utah. Situated close to an abundant supply of pine timber, from which lumber, shingles, etc., could be cheaply manufactured, building was comparatively easy. During the immense emigration of former years it had been the favorite recruiting place for people en route to California; and many here lingered for a few days or weeks to rest their cattle, and lay in stores preparatory to surmounting the last, then formidable, barrier—the Sierra Nevada Mountains—which barred the pathway to the Mecca of their weary pilgrimage from the far East; and it bid fair to become an important frontier town, which would be able to nourish upon its own natural resources. The discovery of the famous silver mine a few miles to the eastward of this prospective inland city, however, effected an entire change of circumstances, and, consequently, a change of futurity awaited it—a different history than that anticipated by its early residents.

MAILS, STAGES AND EXPRESS.

Colonel A. Woodard and Mr. Chorpening had associated themselves together, and under the firm name of A. Woodard & Co. made a contract with the United States in 1851 to carry the mail from Sacramento, in California, to Salt Lake City. This route, commencing at Sacramento, ran via Folsom to Placerville, in El Dorado County; thence over the Sierra by the old emigrant road, through Strawberry and Hope Valleys into Carson Valley, through Genoa, Carson City, Dayton, Ragtown, and thence across the Forty-Mile Desert to the Humboldt River, near the Humboldt Sink; then following the old emigrant route east along the Humboldt River to what is now Stonehouse Station, on the Central Pacific Railroad, near which it left the river and, turning to the southeast, took the "Hasting's Cutoff" to Salt Lake City. The entire length of this route was 750 miles. The mail was packed on the back of a mule, and the trip was made once a month each way.

Genoa Stage Lines crowded with 15 rather dapper looking men onto its open "Jupiter" wagon.

Steve Crandell Collection

California Oct. 20th 1849

Beloved Companion

One strange peculiarity prevails here for modern times. Every man is allowed as many wives as he can support. Bah! (with more rather have the undivided affections of my one than share it)

I have not yet been to the Post Office to enquire for that letter, for I do not expect it is there The U.S. mail is expected in about two weeks, which will just be in time to leave me without one solitary word of information from you until I get away through to California. But I hope there to find enough to compensate for all. And all I can do is to wish you in the cheerful enjoyment of Heavens richest blessings Remember me to all the Family, and enquiring Friends

And believe me your devoted and Affectionate Husband

E. A. Spooner

P.S. Lost my gold pen and have to write with a crow quill, alias Turkey Buzzards The prospects for obtaining gold are not as good as we had

Galatt's Livery Stable, Genoa, early 1890s

After travelling about 60 miles we entered Carson Valley, proper.

This is a lovely valley about 10 miles wide and 30 long of exceeding fertility with two fine streams flowing through it viz the North and South forks of the Carson. But a few acres have as yet been settled on, but a town is forming at the Mormon Station about half way up the valley.

William Henry Hart, September 6, 1852

The actual difficulties to be surmounted, and the dangers, real and fancied, that beset the whole line, are too numerous to recount, and beyond the powers of imagination to correctly paint. In the winter, upon that portion of the route which passes over the Sierra, the snow fell from fifteen to twenty feet on a level, and in the cañons and mountain gorges drifted to the depth of forty or fifty feet. In the spring the Carson and Humboldt Valleys were sometimes flooded, and swimming was the only means of passage, as there were no bridges. From Stone-house Station, east, the whole country was infested by bands of hostile Indians. The Shoshone tribes were the worst, and gave the most trouble. They would skulk behind the rocks and watch day and night for the mail or emigrant train, lying in wait to kill and plunder. So great were the dangers from this source that it "was found necessary to employ men to travel with and guard the mail. In the fall of 1851, Colonel Wood-ard, while in charge of the mail, and two young men, John Hawthorn and Oscar Fitzer, who were employed as guards, encountered a band of these hostile tribes at Gravel Point, near Stone-house Station, and were all three killed. Chorpening, the surviving partner, continued to carry the mail till the fall of 1853, when this contract expired. He was then joined by Ben. Holliday, and they obtained permission to carry the same with a four-mule team and covered wagons, which they continued till June, 1857, when the establishment of a tri-weekly line of stages from Placerville to Genoa, by J. B. Crandall, left them with the line only between Genoa and Salt Lake.

PIONEER STAGE LINE. In the summer of 1857, Col. J. B. Crandall established a tri-weekly line of stages between Placerville and Genoa, and carried the "Carson Valley express," which was managed by Theodore F. Tracy. E. W. Tracy was agent at Placerville, and Smith and Major Ormsby were agents at Genoa. In June of that year [he] established the following stations between Placerville and Genoa, viz.: Sportsman's Hall, Brockliss Bridge, Silver Creek, and Gary's Mill. This was called the "Pioneer Stage Line," and connected at Genoa with the Chorpening wagons to Salt Lake.

OVERLAND MAIL. The summer of 1858 marked a new era in mail and stage facilities. Crandall transferred the Pioneer Stage Line to Lewis Brady & Co., who established a semi-weekly stage between Sacramento and Genoa. Major George Chorpening, brother of the enterprising and indomitable stage proprietor, had secured the United States mail contract from Placerville to Salt Lake City, which was to connect at that point with the regular overland mail to St. Joseph, Missouri. This put new life into the route from Carson to Salt Lake, and raised fresh hopes for the future of the region of country along its line. The first coach under this arrangement left Placerville June 5, 1858. The first Overland mail stage, bringing letters and passengers from the East, arrived in Placerville, Monday,

At this camp along the Carson River in 1854 this flaxen-haired young woman was doing the unthinkable . . . letting her hair down. It was rare to find a woman without a bonnet in this era. Perhaps it's this brazen behavior that's the cause of the twinkle in these young men's eyes.

Steve Crandall Collection

This had been a most disastrous piece of road to those who had preceded us. The sand was of sufficient depth to cover the wagon fellies as the jaded and worn-out animals labored under the stimulant of the brad and lash to draw their burdens. "It was the last straw that broke the camel's back." The last 10 miles we could walk almost the entire distance upon the bodies of dead and dying animals, horses, mules and oxen, by the score, still attached to the wagons, lying in and along the roadside, in harness and yoke. Drivers, with women and children, had abandoned all to seek water and save their own lives. The stock with sufficient strength left to travel in some instances were detached from wagons and urged along, loose, before them. The ground was strewn with guns, ox chains and every kind of thing that had been abandoned. And to this day that sandy plain is covered with the bleached bones of the faithful beasts that perished on that fatal desert. By exercising due care and caution, I passed over the ground in safety with my train in 1853, with all the evidences of the terrible losses in '49 and '50 still visible.

David Augustus Shaw, 1853

Galatt's Livery Stable, Genoa, late 1890s. If you look at the last picture of this stable you can see that an addition was put on this building.

PETRIFIED MAN

A petrified man was found some time ago in the mountains south of Gravelly Ford. Every limb and feature of the stony mummy was perfect, not even excepting the left leg, which has evidently been a wooden one during the lifetime of the owner - which lifetime, by the way, came to a close about a century ago, in the opinion of a savan who has examined the defunct. The body was in a sitting posture, and leaning against a huge mass of croppings; the attitude was pensive, the right thumb resting against the side of the nose; the left thumb partially supported the chin, the forefinger pressing the inner corner of the left eye and drawing it partly open; the right eye was closed, and the fingers of the right hand spread apart. This strange freak of nature created a profound sensation in the vicinity, and our informant states that by request, Justice Sewell or Sowell, of Humboldt City, at once proceeded to the spot and held an inquest on the body. The verdict of the

continued in far right column

July 19th of that year, at ten o'clock in the evening. The event caused universal rejoicing, and was celebrated with bonfires, speeches and other demonstrations of joy and gladness.

Many difficulties and dangers attending its passage made it necessary to send special messengers a portion of the way to guard. Messrs. Rightmire and Lindsay, most worthy and efficient gentlemen, were employed to accompany the mail-coaches as far as the Big Meadows, near the Sink of the Humboldt, and return with the westward bound stage. On their return, July 13, 1858, they reported having met, on the third of July, five emigrants who came through from Iowa that season, at the Sink of the Humboldt, who took the Truckee route for California. They had crossed the country on pack-mules, and according to a report published in the Mountain Democrat of Placerville, at that date, they overtook General Harney and troops on the Sweetwater in the Rocky Mountains, en route for Salt Lake City who gave them peremtory orders not to pass through the Mormon country, which they had complied with by going to the north of the City of the Saints. They further stated that in Hot Spring Valley they overtook a train consisting of sixteen Mormon families (most of whom were women), hastening on to Carson Valley. These families were, they said, in perpetual dread of being pursued and massacred by the Salt Lake Mormons, and were making almost superhuman efforts to widen the distance between themselves and the sanguinary saints.

On the fifth of September, of the same year, Mr. Lindsay returned with the overland mail-coach, having a portion of the Salt Lake mail of August 16th, also the mail which left there August 23d. He reported an attack upon the mail party, August 20th, by the Shoshone Indians, and the destruction of their wagon and part of the mail matter. It appears, from the account given at the time by the Mountain Democrat, that on the night of August 20th, while encamped eight miles below the first crossing of the Humboldt, the mail party of August 16th were surrounded by a large body of Shoshone Indians, who, by yelling and hooting, succeeded in stampeding and driving off the stage animals. Mayfield, the conductor, and his assistants, remained during the night to guard the wagon, but in the morning, finding that the Indians had gathered in great numbers, they determined to abandon everything except their arms and ammunition, and take to the mountains for personal safety. The mail-coach was afterwards found, literally torn to atoms.

In January, 1859, the overland stage brought the President's message from Salt Lake in seventeen days. Letters sent by the overland mail reached their destination in the East ten days in advance of the ocean steamer, and as a stage left once a week this line began to be the more popular and more generally patronized by the public.

In October, 1859, J. A. Thompson and Judge Child started a new stage line to run tri-weekly between Placerville and Genoa. They run with coaches from Placerville to Strawberry valley, and from there to Carson Valley they used sleighs, and thus kept the line open all winter. For this purpose they

Sheridan Blacksmith Shop, Genoa

Courtsey of the Nevada Historical Society

continued from far left column

jury was that "deceased came to his death from protracted exposure," etc. The people of the neighborhood volunteered to bury the poor unfortunate, and were even anxious to do so; but it was discovered, when they attempted to remove him, that the water which had dripped upon him for ages from the crag above, had coursed down his back and deposited a limestone sediment under him which had glued him to the bed rock upon which he sat, as with a cement of adamant, and Judge S. refused to allow the charitable citizens to blast him from his position. The opinion expressed by his Honor that such a course would be little less than sacrilege, was eminently just and proper. Everybody goes to see the stone man, as many as three hundred having visited the hardened creature during the past five or six weeks.

Mark Twain, for the *Territorial Enterprise*, October 4, 1862

Christensen Homestead, Genoa, Nevada

After journeying for two or three miles, we found there was plenty to try the temper of the passengers. We began to feel cramped, the heat of the sun made us hot and irritable: and not only was there a difficulty about stowing away one's feet, but we had even to fit in our knees one with another, and then occasionally give and take pretty smart blows caused by the jostling of the carriage. Most of the men chewed tobacco, and those who occupied centre seats had to exert considerable skill to spit clear of the other passengers. Americans are generally adepts in this art, but we had one or two unskilful professors, although it must be admitted that they had hardly a fair opportunity of showing off their proficiency, from the jolting of the coach.

Edmund Hope Verney, 1865

built two fine sleighs, with three seats each, in December, 1859, which were the first sleighs ever used on this mountain road. In the spring of 1860 Louis McLane purchased the "Pioneer Stage Line" between Placerville and Genoa, which he transferred in the year 1861 to Wells, Fargo & Co., who then run the entire route to Salt Lake. In the summer of 1860 A. J. Rhodes started an opposition stage line between Placerville and Carson City via Genoa. He used six-horse coaches, made daily trips in from ten to twelve hours and reduced the fare from forty dollars to twenty dollars. In the summer of 1862 he sold out to McLane, binding himself not to start another opposition line.

TELEGRAPH LINES. The first movement towards an Overland Telegraph line was made at Placerville in 1858, by the organization of the Placerville and Humboldt Telegraph Company. The first pole was erected at Placerville July 4, 1858, and the line built to Genoa that fall, and extended to Carson City in the spring of 1859, and to Virginia City in 1860. It was not completed to Salt Lake till the fall of 1861.

PONY EXPRESS. In the spring of 1860 the celebrated Pony Express was established by Jones, Russel & Co. W. W. Finney as agent, organized the line between Sacramento and Salt Lake. The express came from San Francisco by steamer to Sacramento, and was there immediately taken by a man on horseback. The old emigrant route was followed across the Sierra till the valley of the Carson was reached, when the Simpson route was adopted. This led to the east, through the desert in Churchill

County, crossing the Reese River at Jacobsville; thence northeast to Ruby Valley and thence southeast, passing out through Deep Creek and around the south end of Great Salt Lake to Salt Lake City. The time between Sacramento and Salt Lake by the Pony Express was three and one-half days — relay stations every twenty-five miles. One rider covered seventy-five miles, and he was given but two minutes at each station passed. The average rate of travel was nine miles per hour. The schedule time from New York to San Francisco was thirteen days, via St. Joseph, Missouri. The first express left Sacramento April 4, 1860, at 2:45 p. m., and carried fifty-six letters from San Francisco, thirteen from Sacramento, and one from Placerville at five dollars per letter. The first express from New York arrived April 13. 1860, bringing eight letters. The time from St. Joseph was ten days. The third trip of the express brought news of the result of the prize fight in London between Heenan and Sayers. Also of the adjournment of the Democratic National Convention at Charleston, South Carolina, to meet in Baltimore the eighteenth of June following, as there had been no agreement upon a Presidential candidate. The quickest time on record made by the Pony Express was with President Lincoln's first message. The time taken in bringing it from St. Joseph, Missouri, to Carson City, a distance of 1,780 miles, was five days and eighteen hours. It was done with double sets of horses, i. e., with fresh horses between stations.

Most all of the cities in the west built up and burned down -- often several times in a span of a few years. Logging was a huge business with never enough lumber to meet the need. Where there was no water to float the logs to the mills, wagons and mule or ox teams would often haul the heavy load miles to the mills.

We bid good bye to our friends with the oxen and pushed on to the famous Kanyon. Found it a much worse place than we had ever anticipated. For Six miles the road was over and among large and small rocks, trees, quagmires, steep pitches +c possessing all the qualities which go to make up a most horrible and all but impossible road.

Many cattle are lamed and wagons broken up in this awful Kanyon. It is, the gap in the eastern range of the Sierra Nevadas made by the North Fork of the Carson in its flow from its high mountain sources, to the valley.

William Henry Hart, September 6, 1852

The excitement was high for the short-lived services of the Pony Express. This photo is believed to be the only known photo of an actual rider taken during the Pony Express' heyday. Experts tell us though that between his dress, his pack horse and the tack and gear he's carrying, that he's clearly not riding for the Pony Express the day this picture was taken.

The severe discomforts of this travelling [by stage] can hardly be exaggerated, but one learns to endure them. The character, the language, and the manners of the class of people who chiefly use this route, however, became if possible even more repugnant to me each day. These I could not endure without disgust.

Edmund Hope Verney, 1865

GENOA OF THE PRESENT DAY [1881]. The little hamlet is busy, with no excess of population, and consequently there are no idlers nor tramps.

Of saloons, where but in a frontier town of equal population would one think of finding six?—the number in Genoa. This preponderance of drinking-places of itself indicates that the present residents are a social, jolly, bibulous class.

THE COURT HOUSE BUILDING. Genoa, being the shire town of Douglas County, also numbers among its architectural features a fine Court House. This building was erected in 1865; is of brick, with iron doors and shutters, and intended to be fire-proof.

The Nevada and California Telegraph Company has its main office here. This is a private enterprise, and was completed in September, 1878, in circuit with Virginia, Gold Hill, Carson, Silver City, Dayton, and Empire. It communicates direct with the Mountain House and Aurora, in this State, and with Colville, Bridgeport and Bodie, California. Length of line 112 miles. It works direct with Virginia on San Francisco business, messages being repeated from Virginia.

The Genoa Flume and Lumber Company's V flume terminates here, discharging the wood, which is cut high up in the mountains, into the Carson River, whence it is floated to the mills at Empire, or taken from the stream above that place, and hauled to Carson City.

EARLY SETTLEMENT OF GLENBROOK.

Glenbrook is located in a beautiful cove on the shore of Lake Tahoe, and is the great lumber manufacturing town of the State of Nevada. The site of Glenbrook was first claimed and squatted upon in the Spring of 1860, by G. W. Warren, N. E. Murdock and R. Walton.

In 1861 Capt. A. W. Pray erected a saw-mill, which was for several years run by water, conducted through flume and ditch for more than half a mile, but the constantly increasing demand for lumber, and a lack of water in the dry portion of the year, compelled him to abandon that motor and resort to steam. This he did in 1864, the newly modeled mill costing $20,000. These were the first mills built upon the soil of Nevada, at Lake Tahoe, though one had been constructed in Lake Valley, California, in 1860, now known as Woodburn's Mill. At the begin-

Some time late in the fall of this year [1864] a young man named White, who had previously lived at Genoa, but more recently kept a wayside inn at New Pass, having had some difficulty with his wife, seized their child of a year old, and started with it in his arms, on horseback, across Reese River Valley. Friends of the wife pursued White for the purpose of taking the child away from him; and when about to overtake him, near the old town of Clifton, Lander County, he placed a revolver at his child's head, and blew out its brains. Then turning the pistol to his own head, he again fired, and fell from his horse a corpse, thus completing the horrid tragedy.

Hanson's Saloon in Genoa

Courtsey of the Nevada Historical Society

Glenbrook Bay

September 15.[1870] ——— Lyon was killed by James Stuart, at Glenbrook Hotel, Lake Tahoe. Lyon was the aggressor, and repeatedly attempted to cut Stuart, when the latter stabbed him fatally.

ning of the enterprise Captain Pray had partners, but he eventually bought out their interests, and in 1862 also purchased the possessory title of the original locators—Warren, Murdock and Walton—at a cost of about 89,000. The old pioneer mill is yet standing. With far-seeing sagacity Captain Pray secured from the Government a title to 1,000 acres of land—locating it with Sioux scrip. A portion of this land was heavily timbered, while some was excellent for grain, hay and vegetables.

PRODUCTIVENESS OF THE SOIL. The productiveness of the soil upon the lake shore is somewhat wonderful, considering the rigorous winters and its high altitude. But the soil, being the fine debris from abraded granite, very soon warms up under the influence of the summer sun after the disappearance of the snow. Captain Pray has several hundred acres under cultivation, and it is no uncommon thing to cut four tons of timothy and, clover hay to the acre, while three tons are a certainty. The hay land is not irrigated. The average of the wild hay crop, or indigenous grasses, is about one and one-half tons per acre. Wheat and barley grow profusely, and Captain Pray's crop was so large one season that he brought in a reaper to harvest it. He thinks he has harvested some crops that have yielded sixty bushels of wheat to the acre, and there have been instances where oats have been measured that stood seven feet and eight inches high. Hay, baled for the use of logging teams, sells at Glenbrook at twenty-five to thirty dollars per ton.

FRIDAY'S STATION AND THE NEW ROAD. In 1860 J. W. Small and M. K. Burke located the place a few miles above Glenbrook, upon the Placerville road, and built the house which has ever since been known as "Friday's" Station. This is about three-fourths of a mile inside of the Nevada State line, and Mr. Small still lives there. At that time all the travel, which was becoming very great, entered Carson Valley principally by the way of the Kingsbury Grade. In 1862 a new route was contemplated from Friday's Station to Carson City, following the lake shore for some distance, and then diverging into the head of King's Cañon, and entering Eagle Valley at the Capital City. This road was called the Lake Bigler Toll road, was of easy grade for a mountain thoroughfare, somewhat shortened the distance to the great mining center, and was completed in 1863.

FIRST HOTEL AT GLENBROOK. The new road diverted much of the travel, and, consequently, eligible sites for public houses were sought along its line, and buildings for this purpose erected. Of these there were none more suitable nor pleasant

Tahoe's first resort was built in 1863 at Glenbrook to provide a luxury vacation spot close to the Nevada mines and the head of the new turnpike to Carson City. It was an elegant resort amidst beautiful surroundings.

Special Collections, University of Nevada-Reno Library

I noticed a single pedestrian coming at no great distance. When he came near I was surprised to see Orin Moody, one of our party. He was without coat, vest, blanket or any incumbrance whatever. He said he was sick and looked it. He was the individual who took my last drop of water on the desert. The first thing he said, after mutual greetings was, "For heaven's sake, have you anything to eat? I haven't had a bite in 24 hours. I took from my pack the remnants of my roast spare-ribs and an "emigrant biscuit"--a cold pancake--and passed them to him. He sat by the roadside and eagerly devoured them. Upon inquiring what had become of the rest of our company, he replied he did not know. He became lost from them the previous morning, having started to walk along before the others were ready to leave with the pack animals. Upon realizing his situation and giving up all hope of joining the company, his only safety depending upon overtaking me. He had become exhausted and ill by his long, rapid walking, and was overcome by heat and hunger. He declared he could walk no further.

I arranged my pack behind the saddle, and helped him to mount my pony. After traveling a few miles his condition compelled us to stop at the first convenient spot. I spread my blankets and he lay down in the partial shade of a few willows in a state of perfect collapse. He beggged me to go and leave him to his fate. I replied that whatever was to come we would share it together, no matter what the "fate" might be.

David Augustus Shaw, 1850

Glenbrook was a vibrant location on Lake Tahoe with resort guests and loggers sharing this beutiful lakeside location.

...an incident occured: both startling and instructing. Mr Martin, Scott of Peoria Ill (our traveling Companion) was struck by one of the animals throwing its head around: he was brought to the ground with a heavey fall. I saw him fall, and started to his relief: but as he began to struggle as if in the last agonies of life, and not fully knowin the cause. I suppose it might be a stroke of: or an arrow from an Indians quiver, I paused; thinking the liveing might need more assistance than, those already dead I viewed "The Camp" with a quick glance, and no frightful met my view. And as Scott began to show more signs of life I haseteened to him: by this time some of the boys came up. We helped him to his feet - his senses were gone; and soon as he could stand; he threw us from him frantic with rage. It was with the up most difficulty we could dissuade him from the belief, that some of us had struck him. This was a very startling scene from its severity, and attending circumstances. Instructive as it caused us to weigh the importance of a sane mind.

John Furmes Cobbey, 1850

than the little cove upon which Fray's mill was located, and the same year that the road was finished, 1863, Winters & Golbath erected the large structure which has since been known as the Glenbrook Hotel. This property now belongs to Yerington & Bliss.

MORE MILLS AT GLENBROOK. Lumber was at this time in good demand, and a common article readily commanded twenty-five dollars per thousand, and clear, forty five dollars. No one person could be expected to long maintain the exclusive monopoly of its manufacture, and the Pray Mill was followed by one erected by J. H. F. Goff and George Morrill in the north part of the town. This did a good business until it was destroyed by fire. The site and remaining plant were then sold to A. H. Davis & Son who built the Davis Mill, which eventually passed into the possession of Wells, Fargo & Co., and is still retained by that firm.

In 1873 the firm of Yerington & Bliss began the lumber manufacturing business at Glenbrook, and have since that time been among the heaviest dealers in the Pacific Coast States. They have connected the timber belt of the entire valley of Lake Tahoe, as well as the surrounding mountains sloping toward it, by rail and V flume with the trans-mountain valleys and great consuming points of the interior. During the year 1873, at the time that firm began operations, the lumber product of

Douglas County was only 906,000 feet. This rose in 1875 to 21,700,000 feet, but with the enormous consumption of timber it had fallen in 1880 to 12,000,000 feet.

FIRST STORE AT GLENBROOK. The lumbering, dairy and other interests, which were springing up all around the lake, soon attracted quite a population and created the necessity of establishing a trading-post at a point best located for a general distributing depot. Glenbrook was selected as the most available spot, and in 1874 J. A. Rigby and A. Childers built the first store, and offered for sale the first stock of goods there. The building was built over the water, and set on piles in front of the present hotel. This may in some degree account for the mysterious disappearance of Mr. Childers, who came up missing one morning, and was never heard of more. It is surmised that he may have accidentally stepped off into the water and been drowned, as from this peculiar sheet of water the body of not a single person drowned therein has ever yet been taken. After the disappearance of his partner, Mr. Rigby admitted into the business W. T. and S. C. Davis, and the firm name was then changed to Davis Co, & Bro. In January, 1877, the building was burned, and the same year Captain Pray built the present handsome two-story structure, 30x62 feet, the corner part of which is now occupied as a store by J. M. Short, and the upper story as a hall.

Mill at Spooner's Summit, 1876. The voracious need for lumber had mills working day and night, clear-cutting away the beautiful forests the area is known for today.

Special Collections, University of Nevada-Reno Library

When they have occasion to punish a man for any petit offense, such as murder in the second degree, women stealing, kidnapping, or such like, they confine him in some dark, lonesome place, deprive him of his accustomed beverages, so that his addled senses return, and he has time to think. Sometimes it is very difficult for him to escape from these places, still if he does not succeed in getting away, they usually give him his liberty for his pains, as the trouble of using his brains and exercising his faculties in making an escape is considered a sufficient punishment for most light offences. In former times they frequently resorted to the death penalty. Sometimes this penalty was inflicted because a man killed his brother, possessed too much money, and sometimes for borrowing other men's wives. As there are no natural projections in this country by which a man can be hanged by the neck, as done in other places, they tie his feet, draw a sack over his head, then throw him into some mining shaft fifteen or eighteen hundred feet deep. Here the prisoner is left to grope about in the dark until he perishes from loneliness and thirst.

Caroline M. Nichols Churchill, 1874

S.S. Tahoe, Glenbrook, Nevada, was the longest and grandest ship ever to grace the waters of Lake Tahoe. It was constructed in San Francisco in 1894, then disassembled and transported in sections by train and horse drawn wagons to the lakeshore at Glenbrook, Nevada. The 169 foot steel-hulled steamship was reassembled and launched on June 24, 1896. Outfitted in an elegance befitting the "Gay '90s," the steamer had polished brass fittings, a teak and mahogany trimmed deckhouse, leather upholstery, hand-woven carpeting, and marble lavatory fixtures.

Having heard such alarming accounts of the difficulties of the Carson and our wagon being in need of repairs and one of our cattle tenderfooted we concluded here to sell our team and buy horses or mules to cross the mountains with. Consequently we sold the team (4 yoke) and wagon for $410 and disposed of such utensils and provisions as we had no further use for and bought horses and packed them with our clothing + provisions, utensils +c. My horse and saddle cost over $65 so that after repaying to [Thomas] Russell the $30 I borrowed of him in Salt Lake I had about $40 left.

William Henry Hart, September 6, 1852

THE VILLAGE OF GLENBROOK. Besides the Glenbrook Hotel, in the spring of 1876, Captain Pray converted his planing-mill, which had been used in conjunction with his saw-mill, into a hotel, which he christened the Lake Shore House. There are also several boarding-houses in the village. Glenbrook supports two saloons, both being conducted by the same firm, however, B. Greenhood and Levi Knowles. The first saloon in the place was opened by Rice & Comstock, in 1877, in John Toll's building.

There are at Glenbrook thirty cottages, one saw-mill, one hotel, one store, one saloon, one livery stable and one meat market, all under one ownership, that of Capt. A. W. Pray. These rent as follows: Cottages, from five to ten dollars per month; the store for $1,072 a year; livery stable, twenty-five dollars per month; meat market, fifty dollars, and the hotel for seventy-five dollars a month.

The cottages are principally rented to the employes of the mills, those engaged upon the railroad and flumes, and the wood-choppers, with their families. An occasional tourist stays here for two or three of the hottest months of the year, and there are quite a number of transient pleasure-seekers passing through from stage to boat, but only a few who are not regularly engaged in business tarry long.

THE SHAKSPEARIAN ROCK. A singular illusion is presented to the observer, from Glenbrook—the profile of a man reclining, with face upturned, appears at the apex of a mountain peak. From a fancied resemblance to the greatest of all poets, it is called Shakspeare Rock. It was first noticed in 1862 by the wife of Rev. J. A. Benton, of California, who was at that time sketching the mountains.

A MOST DEPLORABLE AFFAIR. Near Shakspeare Rock is a cavern, the entrance to which is ten feet high and twenty wide, upon going through which to the north the cavern is reached, being about twenty feet wide, seventy-five feet long, and about fifty feet high to the roof. To reach this, ropes are required, and great caution must be observed lest by a false movement the explorer be plunged into the yawning chasm below. It was to view this place that a gay party of young people from Carson City left the village of Glenbrook on the ninth of September, 1877. The party were Miss Carrie Rice, Miss Vade Phillips, Miss Esther Moody, Frank Williams, William Clark, and William Cramer, the latter being Miss Rice's escort. "Upon reaching a

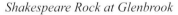

Shakespeare Rock at Glenbrook

Elder Hyde started from Mormon Station late in October, with a companion named Willis Lewis, to cross the mountains to Placerville, California, to procure machinery. They were caught in a severe snow-storm, and Lewis declined to proceed, and endeavored to return. He was never heard of again, and his bones, washed by the driving rains and covered by the drifting snows, no doubt lie in some secluded spot in the beautiful Sierra, whose towering peaks reveal not the many tragedies they have witnessed. After a desperate struggle to free himself from the encompassing snow, Elder Hyde finally reached the station completely exhausted, his feet frozen so badly that their preservation was despaired of. For several months he went upon crutches, and it was a long time before he fully recovered from the effects of the terrible exposure that had been fatal to his young companion, and had brought him so near the brink of death.

Sam Brown selected his "man for breakfast" from the class without friends, and then deliberately murdered him only when he knew perfectly well that his own person was safe from harm. He was an arrant coward, and did his killing mainly when he had been stimulated to courage by strong drink.

Henry Vansickle, a genial man withal, and a quiet citizen, lived three miles above Genoa, where he kept a hotel. Thither rode Brown and his companion, arriving there just as the bell was ringing for supper. Here thought Brown, is a man who will not fight, and can be safely killed. Brown dismounted from his horse, and was in the act of untying his leggings. Vansickle then asked him if he should put up his horse. Brown then turned to Vansickle and in his rough way said: "No, I would not stop with you, but I have come to kill you, you ——!" and immediately drew his pistol. Mr. Vansickle was taken completely by surprise, and was not armed. He was well acquainted with Brown, from his often having stopped at his hotel, had never had any quarrel with him, and Brown had never before exhibited any ill-feeling towards him. As quickly as possible, and before Brown could shoot, Vansickle rushed into the dining-room, at that time filled with guests at the supper table, Brown following, pistol in hand. Instinctively the persons at the table all jumped to their feet, thus covering Vansickle's retreat effectually. Without shooting, Brown then went out of the house and rode off up the road. Vansickle

continued in far right column

Henry Van Sickles hotel, stable & corral, Carson Valley

precipitous point overlooking the cavern, where the surface presented but a smooth, solid granite front, and where the entire party should have paused, these two young people, Mr. Cramer and Miss Rice, clasped hands, and thoughtlessly started down the inclined plane leading to the chasm, thinking they could stop upon its brink. Swiftly they shot down toward the fateful precipice, and when too late, essayed to check their furious progress. Both fell down. Miss Rice went over the precipice, and when aid was procured was found lying about ten yards from the mouth of the cavern, breathing, but unconscious. She died in an hour. Her escort fortunately succeeded in grasping something to which he held, and was rescued.

RAFTING LOGS ACROSS LAKE TAHOE. As the forests in the immediate vicinity of Glenbrook were denuded of timber, the millers were compelled either to suspend operations or draw upon some other source for a supply of logs. On the western shore of Lake Tahoe, in California, were virgin forests of immense trees, extending from the water's edge, upon the sloping foot-hills, to the deep snow line on the mountain sides; and human ingenuity sought and soon adopted methods to render this large reserve available. Steamers were brought into requisition; the trees were felled, cut into suitable lengths, hauled upon trucks drawn by oxen and rolled into the water. The logs are then confined in "booms," consisting of a number of long, slim spars fastened together at the ends with chains, which completely encircle a "raft " of logs arranged in the form of a V—some of these rafts containing timber enough to make 250,000 or 300,000 feet of lumber, In this condition they are attached to the steamer with a strong cable, and towed to the mills at Glenbrook, which, being built immediately upon the lake shore, are so arranged that the logs can be hauled by machinery upon the ways to the saw carriage as required.

A number of small steamers are employed for this purpose; and the noble forests that once were the pride and beauty of the western shore of Lake Tahoe are fast disappearing before the destructive ax of the woodmen, and they, too, will soon be a thing of the past. The principal vessel used at this time for towing logs is an iron tug called the Meteor. This boat was built at Wilmington, Delaware, by Harlan Hollingsworth & Co.; after having been put together it was taken down, shipped by rail to Carson City, and then hauled to Lake Tahoe by teams. This was in 1876. The Meteor is a propeller, eighty feet long and ten feet beam, and will run twenty miles an hour under a pressure of 135 pounds of steam. This vessel cost $18,000, and when not engaged in towing logs, is frequently seen making the tour of the lake with some distinguished personage on board.

Man's best friend watches while a boom of logs is moved on Lake Tahoe.

Special Collections, University of Nevada-Reno Library

continued from far left column

in the meantime had got possession of his gun—a double-barreled fowling-piece—and taking in the situation, concluded that as Brown had begun upon him he might as well settle the affair at once, and not live in fear of future attacks. The gun was loaded with fine shot, which Vansickle did not take time to draw, but added a charge of buckshot to each barrel. Then, having ordered out a horse, saddled, he mounted the animal and gave chase to Brown—an avenging Nemesis. Overtaking Brown and his companion about a mile up the road, and when getting within shooting distance, Vansickle called to Hendereon to get out of the way, which he quickly did. Vansickle then discharged both barrels of his gun at Brown, knocking him off of his horse, but not seriously wounding him, for he soon remounted and fired two shots from his pistol at Vansickle, and then rode on as fast as he could. Vansickle followed him with his empty gun until he arrived at Mr. William Cosser's house in which Brown had taken refuge.

Meantime, several persons had followed Vansickle from his house, and here overtook him, who had been thoughtful enough to bring with them a supply of ammunition, with which Vansickle again loaded his gun. Brown, soon after, came out of the house and started up the road in the direction of Olds' Station, with Vansickle in pursuit. Having the fleetest horse, Vansickle overtook

continued next page, far left column

continued from last page, far right column

Brown, near Mottsville, and again discharged both barrels of his gun, but without apparent effect. Brown then turned and, after firing three shots at Vansickle, rode up to the residence of Mrs. Mott and took refuge in the house. By this time it began to grow quite dark, and Vansickle, not caring to attack his enemy while he was under cover, watched the premises until be should come out. After waiting for some time and seeing no appearance of Brown, and a person happening along the road, Vansickle prevailed upon this passer-by to enter the house, and report whether or not the bird had flown. This man reported that Brown was not there. Whereupon Vansickle hurried on to Luther Olds' hotel, expecting to find Brown there ahead of him. But be was disappointed— he was not there. He remained there, however, for a short time, and at length heard the jingle of spurs which he recognized as those worn by Brown. Immediately leaving the house, Vansickle reached the road just as Brown had alighted from his horse. Confronting him with the remark, "Bam, I have got you now!" he discharged both barrels of his gun into his breast. Upon seeing his pursuer, mortal terror seized upon the ruffian; abject, unutterable fear sealed his lips; a spasmodic, agonizing yell of despair involuntarily forced itself from his mouth, "piercing the night's dull ear," and the inhuman monster was dead!

Walley's Hot Springs in Genoa

MEDICINAL SPRINGS OF THE COUNTY.

Near Henry Vansickle's, at the base of the mountain spurs which jut into the valley from the west, two miles south of Genoa, are some large thermal springs, now known as Walley's Hot Springs. Here is a large hotel building containing forty rooms, with bath-houses adjoining. There are eighty acres of land belonging to the property, and the improvements were made at a cost of $100,000. These springs have a great reputation for the cure of rheumatic and scrofulous affections, and have become a noted place of resort. The location is extremely pleasant, the scenery grand, and the climate in summer invigorating and healthful.

Upon the land of Captain Pray, near Glenbrook, on Lake Tahoe, there is a mineral spring, the curative properties of whose waters in certain complaints is highly lauded. Iron seems to enter largely into its composition.

Carson City in 1876. A political banner hangs over the street for Democrat Samuel Tilden with Vice Presidential running mate Thomas Hendricks. As we know, Rutherford B. Hayes won the election. Until the Bush - Gore election in 2000, the Hayes election held the honors as the most fiercely disputed election in American history with what was considered a slim margin of 264,000 popular votes.

Chapter 3: Ormsby County - Carson City and vicinity

Of the early trappers and explorers, Kit Carson has left his name applied to the beautiful river that first greets the thirsty traveler from the East and points the way to the crossing of the Sierra, and of the early settlers, Ormsby leaves his name to the county. For many years the white strangers came and went, leaving but their tracks to tell of their passage. Some had

February [1859]. William Bilboa was shot and killed by Sam. Brown, the notorious desperado, in the streets at Carson City. Nothing was done with the murderer, although the act was a wanton butchery.

Steve Crandell Collection

There weren't many women around, but the ones that were there were in great demand. This is a wedding that took place in a camp along the Carson River in 1854.

July 1. Peter Fitzgerald, engineer of the Gould and Curry mine, had a street duel with Sam. Hamilton, a prize-ring sport. After the exchange, of numerous shots, Hamilton was fatally wounded.

tarried a few months, and a few localities in the valley were said to have been "settled," but the great emigration of 1849 — of preceding and later years was for California, and the beautiful valley of the Carson was still a wilderness.

In November, 1851, a party of men from the placer mines of California, seeking gold on the eastern slope, were attracted by the advantages offered for agriculture and trading purposes and located upon ground where now stands the city of Carson. These were Joseph and Frank Barnard, George Follensbee, A. J. Rolling, Frank and W. L. Hall. Killing an eagle on the spot, and preserving the stuffed skin as a trophy, which was used as a sign for their station, the place became known as Eagle Valley. This was the first settlement of the region under review. No government yet threw its protecting aegis over the county. The whole region was a part of Utah.

In 1857 the Mormons were summoned by Brigham Young to Salt Lake, which unreasonable and tyrannical behest the deluded and superstitious devotees of the Church obeyed, and their settlements in western Utah were abandoned or disposed of to any person offering any price. The same scenes and sacrifices enacted here were repeated wherever the Mormon Church had a "stake," in Utah, California or elsewhere. Those of Eagle Valley went with the others, and the region was left with a new element. At this time a new man enters upon the scene. The following sketch of this person was published in the Carson Daily Index of March 20, 1881:—

Soon after a few Mormon families had ranches in Eagle Valley. As these people wore about to remove hence and return to Salt Lake, a man named John Mankin, whom the early settlers designate as an old pirate, mountaineer and frontiersman, purchased for a mere trifle the possessory right and became the owner or claimant of all the valley land lying between Nevers' Lane, extending to the hills north and south, and the now Prison Hot Springs. This man was a widower with four children, one a daughter named Mace, about twelve years old. With him lived also an Ute Indian boy named "Cap." They resided in a cabin then a little northwest of the present town site.

Mankin was a rough, passionate, illiterate fellow; given to quarreling with his neighbors. He was a splendid marksman with his rifle, which was his constant companion, and in his hands a dangerous weapon. His unpopularity caused some of the "boys" to plan a scare for him one night. Among the party were Jim Menifee and Charles Wolfe. They might as well, as they discovered to their own fright, have attempted to catch a weasel asleep. They disappeared behind a log-fenced corral not an instant too soon to escape a bullet. Mankin was a broad-

The thin atmosphere seemed to carry healing to gunshot wounds, and therefore, to simply shoot your adversary through both lungs was a thing not likely to afford you any permanent satisfaction, for he would be nearly certain to be around looking for you within the month, and not with an opera glass, either.

Mark Twain, 1872

The Nevada State Prison (and Warm Springs on right) is located in Carson City. It is one of the oldest prisons still in operation in the United States. Established in 1862 when the Nevada Legislature purchased the Warm Springs Hotel and 20 acres of land for $80,000, the prison has been in continuous operation since. Abraham Curry, who owned the hotel, was appointed as the first warden. Photo between 1862-1867.

Library of Congress, Lawrence & Houseworth Collection

Just as we have fun with our old-time photos today, it appears the concept may have been used much earlier. This 1895 photo taken at the C & C shaft is of visitors to the mine, not miners.

About this time [October, 1864] an exciting scene occurred in the streets of Austin. An unknown man, supposed to be insane, appeared upon the crowded streets, brandishing a glittering axe, cutting all who came within his reach. Three or four men were either killed by him or dangerously wounded. Passing through town he went on down the Clifton Grade, and, in a few hours, his dead body was found in the road. He had been shot. No legal inquiry was ever made as to who killed him. People, however, generally accredited the deed to a sporting man, since deceased, known as "White-headed Boss." For a long time, thereafter, when any person wished to send a thrill of excitement, that would almost result in a panic, among the crowds that thronged the streets of Austin, he had only to raise the cry of "Look out! here comes the man with the axe!"

shouldered man of fifty-four years, so active, that in sport he would run a race with any one in the country, and there were some extraordinarily active men here in those days. The distance of fifty yards would be measured, and Mankin would lie flat upon his face, and at the word would rise and distance all his competitors.

Mankin took a party to the Walker River country on the pretense of showing them rich mining prospects. Once there, he gave them the slip and returned home. For weeks thereafter he kept his gray stallion saddled night and day, ready to escape, fearing the return and attack by the men he had deceived.

At this period the population was exceedingly scarce, it being represented that by collecting all the people in Carson, Washoe and Eagle Valleys, enough would be present to have three sets in a dance. These gatherings usually took place at Dr. King's brewery, which was made a place of public resort. The settlers of Eagle Valley regarded the Eagle Ranch as the central point, and it was long before any other locality bore its specific name. A station was established on the overland road where it touched the river, three and a half miles from Eagle Ranch, which subsequently bore the name of "Dutch Nicks," the usual name for Nicholas Ambrosia, the first-settler, but afterwards changed to Empire City. Families also located at Clear Creek. Mill Creek, and other localities prior to the discovery of the Comstock Lode and the rush of people to Nevada.

RESOURCES.

The wealth and prosperity of Ormsby County are evidences that it possesses resources of an important character. Situated centrally in the most thickly peopled belt of the "Eastern Slope," it derives great profit from the trade and travel its favorable position demands.

A large area, comprising more than 40,000 acres, extending into the Sierra Nevada, was originally heavily timbered, and, although much has been taken, this forest still constitutes an important resource. In connection with this interest are the various small mountain streams, which afford power for manufacturing the forest trees into lumber. These are Clear Creek, Mill Creek and King's Cañon and small streams flowing into Lake Tahoe. The Carson also affords a great water-power, and numerous quartz and saw-mills are propelled by its force. These streams furnish a perpetual power for manufacturing purposes.

The Nut Pine Mountains bear many ledges of gold and silver-bearing quartz, as well as gold in placers. Iron and copper ores are also found in the same range, and a bed of lignite, once mined for coal, exists in El Dorado Cañon. The dearth of water in this region is a serious obstacle to its development.

Eureka Quartz Mill, Carson River

Turned tailor to-day, and cut out for Joseph a pair of buckskin pants. The mines in the cañons at Walker River cannot be worked by reason of heavy frosts and want of water. Times seem dull, but there are plenty of dances; the charge per couple is five dollars. Feed for cattle is getting plenty again.

Harry Fulstone, January 12, 1859

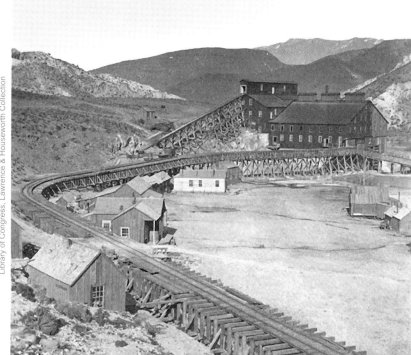

Library of Congress, Lawrence & Houseworth Collection

Carson Street from plaza, Carson City, NV, 1860s

[Regarding Carson City] All is life, bustle and activity at this growing place. Major Ormsby is building an adobe house 45x50 feet, and two stories high. He intends it for a residence and place of business. There is a hotel in progress of construction by Sears & Co., 100x50 feet. Rice & Co., have a large saloon adjoining their hotel nearly completed. Mr. Curry has commenced a building also intended for a saloon. There are also many other buildings in course of construction intended for stores and private dwellings, The scarcity of lumber is a great drawback to our prosperity; J. K. Trumbo disposes of his lumber weeks in advance. Thomas Knott is building a saw-mill in Jack's Valley. A company from Forest City, California, is about building a mill in Eagle Valley and ere long all demands for lumber will be supplied."

Territorial Enterprise, September 17, 1859

The streams of Ormsby, notably the Carson, bearing their freight of lumber, mine timbers and firewood constitute a living and lasting source of wealth. Besides being carriers of the forest products, they afford irrigation for the arid soil, without which there would be no agriculture, no beautiful gardens or shady trees about its dwellings, and more than all, do they afford the power which drives the many quartz and saw-mills which furnish remunerative employment for so large a proportion of the population.

QUARTZ MILLS. The development of the mines discovered in 1859-60 required at once the construction of mills for the reduction of the ores. The first ore extracted was from the Mexican and Ophir claims at Virginia City, and this was packed on mules over the Sierra Nevada to California, some to Grass Valley, and some to San Francisco for reduction, a small portion being reduced in arrastras near the mines. This ore being very rich, one mule carrying $2,000 worth, it was a good enough way of transporting the bullion to market. But there was other ore in the mines not so near pure silver, and this required reducing nearer home. For this purpose the first thought was power, and the "Carson River seemed to offer it in abundance. This stream was about fifteen miles distant, and there at once the enterprising owners of the mines directed their energies.

A small mill was first constructed near Empire City in the spring of 1860, which was subsequently enlarged as the Mexican Mill, or the Silver State Reduction Works. The building of mills once entered upon, the business increased with wonderful rapidity.

In 1861 a mill was built in Clear Creek District and run by water-power from Clear Creek. In the same year a man named Ashe built a mill in Gregory's Cañon, which afterwards took the name of Ashe's Cañon. This mill was destroyed by a flood in the winter of 1861-62 which was so powerful that it reduced the level of the cañon fourteen feet.

SAW-MILLS. The grand forests of the Sierra Nevada were a great attraction to the early settlers of the "eastern slope," offering them facilities for obtaining lumber of which they quickly availed themselves. The first saw-mill erected in the region afterwards embraced in Ormsby County, was built by Mr. Gregory in the fall of 1859, on Mill Creek, three miles west of Carson. This was a steam-power mill, and was the first steam

"Went down to a dance at Jacob's, at Johntown, in Gold Cañon. Walked. Stage overtook me. Sallie King urged me to get on the stage, and I did so. We had a gay time. I came back in Major Ormsby's wagon. It broke down three times, and we had to tie it up with ropes.

Harry Fulstone, March 30, 1859.

Special Collections, University of Nevada-Reno Library

Portrait of a miner in 1895 in Virginia City. From the Consolidated Virginia Mining Company collection. Notice the photo at the bottom left... is it Mom, a girlfriend or a wife?

Carson City stage stop, 1859

November 15 [1858]. You have a deal of trouble here to get your pay after it has been due for months. They are a pack of speculators, robbing one to pay the other. They pay what they please after making agreements, and have it all their own way, and it is of no use to remonstrate.

Harry Fulstone, November 15, 1858

mill of any kind erected in what is now the State of Nevada. The transportation of heavy machinery over the Sierra at that date was a very expensive undertaking, and this was regarded as an enterprise quite extraordinary. The mill was capable of cutting 15,000 feet per day, and for many months was run to its full capacity, so great was the demand for lumber. Orders were taken weeks in advance of the possibility of filling them, and customers contended greedily for their turn.

Shortly after this Mr. Alexander Ashe built a sawmill on Mill Creek near the former, running it by water from the creek. One mile north of Gregory's, Messrs. Thompson & Treadwell erected a powerful steam mill capable of cutting 15,000 feet of lumber per day, also containing a shingle and planing machine, which prepared for market large quantities of material for building purposes. In 1861 these three mills were employing upwards of 100 men; and had cost in their construction $60,000.

Mills now multiplied rapidly, there being in 1862 three on Clear Creek at a distance of from six to eight miles southwest of Carson City. These mills had been erected at an aggregate cost of $33,000. In 1862 they employed 100 men, and were capable of cutting 50,000 feet per day.

The Lake Bigler Lumber Company, C. R. Barrett, A. W. Pray, and K. D. Winters, proprietors, went into operation in 1862 in the region, as the name implies, of Lake Bigler, or Tahoe, where was an abundance of large trees affording a superior quality of clear lumber, compensating for its distance from, and at that time difficult access to market. The mill of the company was propelled by water conducted through a flume and ditch upwards of half a mile in length, giving abundant power. In 1862 this mill contained a set of double circular saws, a muller, edger and shingle saws, employed twelve men and turned out 20,000 feet of lumber daily, besides a large quantity of shingles.

GAME. The word "game" does not, in Nevada, always apply to the animals running wild in forest and field, nor to the untamed birds of the air, neither to the fish of its lakes and streams, although "seeing the elephant " is commonly mentioned, and "hunting the tiger in his jungle " appears to be an every-day, and nightly, sport. A writer of the region says "A man can find there any game he wants, whether played with a pack of cards or pistol; whether it comes in the shape of a big knife, or a straight from the shoulder, or in courtesy and kindness, from the heart." Hunting game, in this acceptation of the term,

Wells Fargo & Company Express and the Ormsby House, Carson Street, Carson City. The Ormsby House was built in 1860 by Major William Ormsby, who was killed later that year in the Pyramid Lake War. The hotel lasted until the early 1900s. Photo 1860s.

Library of Congress, Lawrence & Houseworth Collection

I was prepared to find great changes on the route from Carson to Virginia City. At Empire City—which was nothing but a sage-desert inhabited by Dutch Nick . . . I was quite bewildered with the busy scenes of life and industry. Quartz-mills and saw-mills had completely usurped the valley along the head of the Carson River; . . . clouds of smoke from tall chimneys, and the confused clamor of voices from a busy multitude, reminded one of a manufacturing city. . . From the descent into the cañon through the Devil's Gate, and up the grade to Gold Hill, it is almost a continuous line of quartz-mills, tunnels, dumps, sluices, water-wheels, frame shanties, and grog-shops. Gold Hill itself has swelled into the proportions of a city. Here the evidences of busy enterprise are peculiarly striking. The whole hill is riddled and honey-combed with shafts and tunnels. Engine-houses for hoisting are perched on points apparently inaccessible; quartz-mills of various capacities line the sides of the cañon; the main street is well flanked by brick stores, hotels, express-offices, saloons, restaurants, groggeries, and all those attractive places of resort which go to make up a flourishing milling town. . . A runaway team of horses, charging full tilt down the street, greeted our arrival in a lively and characteristic manner, and came very near capsizing our stage. One man was run over some distance below, and partially crushed; but as somebody was killed nearly every day, such a meagre result afforded no general satisfaction.

J. Ross Browne, 1871

Empire City, 1880

Some of us sought these valleys when they belonged to nature's solitudes, assured that their natural advantages would soon gather society about us. In this we have not been disappointed. The influx of actual settlers has of late been very considerable, and our late holiday frolics should convince an anchorite that society in Carson Valley is a fixed fact. Youth, beauty, intelligence and grace are all here in their freshness and potency, and the spirit of concord seems to preside over our pastimes.

Our New Year's ball at Eagle Valley was a perfect jam. The house, though large, was quite too small. We crowded ourselves out! If any cold-blooded mysogamist doubted the fact that man is gregarious, our New Year's ball would have cured him. All seemed to say in the language of the poet:

"On with the dance, let joy be unconfined."

Territorial Enterprise, January 29, 1859

has often brought "a man for breakfast." But "game" in cities and mining hamlets, and "game" in the open country, in the plains and hills of the broad State, are widely different things.

CARSON CITY. Nestling at the eastern base of the Sierra Nevada is a little valley, nearly circular in form, of about twenty-five square miles of area, separated from the Carson Valley and River, on the south and east, by a low, projecting spur of the Sierra, opening to the river in the northeast, and fronting the hills of the Washoe Mountains in the north.

The people of Carson seemed determined to have a happy time, notwithstanding their many discomforts arising from badly constructed dwellings, the high price of comestibles and the severity of the weather.

The rigors of winter abated about the first of February, giving great relief to stock and their owners, but the deep snow on the Sierra Nevada rendered communication with California exceedingly difficult. With the opening of spring additions were made to the population which had been constantly increasing since the exodus of the Mormons. There is now here the nucleus of a city. The surrounding valley is "claimed" in ranches and occupied by the claimants, herdsmen and station keepers. South is the greater valley of the Carson, with Genoa as its capital, and northeast are Johntown, Gold Cañon and the settlements along the Carson River. A few white men and Chinamen have been washing, or mining, for gold at Johntown and in the cañon at intervals for several years, and now, in the spring of 1859, are meeting with greater success than before. Astonishing developments are made in the mines, and soon their fame spreads abroad. Population flows in, and Carson City has soon grown so large that it would be difficult to keep the record of its individual citizens, although at this date all are pioneers.

August 13th, the telegraph wires were stretched to Carson and an office opened. This was an institution at that time quite uncommon on the Pacific Coast, and the erection of a single line of wire to any town was regarded as an important event.

EMPIRE CITY. Three and a half miles north of Eagle Ranch, now Carson City, the overland emigrant and stage road struck the bank of the Carson River, and there Nicholas Ambrosia located a ranch and kept a station, his claim being recorded March 24, 1855. This station became known as "Dutch Nick's," which name it bore long after the locality had been surveyed into lots and streets, and was officially known as Empire City. The town site was laid out in March, 1860, by Eugene Angel and other surveyors, and the name it now bears given it.

The fine water-power here afforded by the river, and its convenient access to the mines of the Comstock Ledge, were the inducements for making a town. Several large quartz mills were built, as has been mentioned in the history of Ormsby County, and the town has always been busy and prosperous.

Members of the Carson Lodge IOOF, 1890

Courtsey of the Nevada Historical Society

When I was in Carson, a gentleman gave me a quit-claim title to four enterprising black pigs, about half grown. One morning I discovered a hole in a bridge of sufficient capacity to swallow up my pig family, the great fat mother included, and it was near the rendezvous of these Christian animals; I became anxious, and thought it would be wise to complain to the City Fathers before anything serious occurred. I looked in every saloon in town, and at last found them in a basement, contemplating an institution of long-necked bottles, all more or less rheumatic and incapacitated for mending bridges.

Caroline M. Nichols Churchill, 1874

Empire and Imperial Mines

Such things as cutting and shooting are of too frequent occurrence here, and a stop should be put to them. Offenders ought to be placed in confinement until we shall have courts legally organized. It is true some time may elapse before we are blessed with such institutions, but criminals are the persons who should suffer for this delay. They ought to be kept even for forty years, and if they survive the present generation of men and still no courts are organized, we should hand them down prisoners to posterity.

Placerville Observer, a correspondent writing from Carson City, June 26, 1859

Within the town are the Mexican and Morgan Mills, and others in the vicinity. Two miles below is the Brunswick Mill which, when in operation, employs 200 men.

At Empire is the depot of the wood business of the Carson River; the many thousand cords of firewood, mining timber and other classes of lumber floated down that stream are here caught in booms, landed and transferred to the cars of the Virginia and Truckee Railroad which passes through the place, and borne to their destination. Fifty thousand cords of wood were thus brought to market in 1880.

Among the places of business are four saloons and one large store. The present population [1881] is 150.

This is the Wells Fargo Express in Virginia City, probably in the 1860s. Gold mixed with high quality silver ore was recovered in quantities large enough to catch the eye of President Abe Lincoln. He needed the gold and silver to keep the Union solvent during the Civil War. On October 31, 1864, Lincoln made Nevada a state although it did not contain enough people to constitutionally authorize statehood.

Chapter 4: Storey County - Virginia City, Gold Hill and vicinity.

The history of this county is, to some extent, the history of the whole State. It was here that the mines were discovered; here they developed into the wonderful proportions that revolutionized all previous values, and sent trade and manufactures into new channels, built new cities, and sent new millionaires into the world. Though apparently insignificant and unknown

The business men in some localities place their signs upon the tops of their buildings that they may be seen by those occupying a higher position in the world.

Caroline M. Nichols Churchill, 1874

Six Mile Canyon from Virginia City, 1860s

Our road yesterday was pretty much the same. While staying here, a waggon drawn with 4 horses, and with others to relieve them, passed by us. It belonged to a merchant that had in the summer taken some goods to be disposed of in Utah, whom managed, as he told some of the brethren, to make a pretty good business of it. He carried passengers also, of which he had 3, now returning from the Washoe Diggings, having made very good time of it. One of them had a nugget upwards of 1 lb. weight, and numerous smaller ones.

William Ajax, 1861-2

men became fabulously rich and noted, we shall see as our history progresses, that energy and judgment, here as elsewhere, soon asserted their superior values, and gave to their fortunate possessors the control of the great bonanzas. Here, as in all countries and in all times, the presence of great wealth drew together, not only the energetic men of business, but also the criminal and abandoned classes, those who fasten themselves on society, and gather a large share of the produces of the industrious. Gamblers, thieves, swindlers, bummers and prostitutes—all claimed a share of the silver mountain, and, though such people hardly ever retain for any length of time their ill-gotten gains, they manage, somehow, to handle a great share of the money.

DISCOVERY OF THE COMSTOCK LODE. The lode was found in 1859, and a small portion of the community were soon aware of the fact that an important discovery had been made. The few sacks of ores that were shipped to San Francisco were like the few samples of gold that found their way East, which only indicated the vast possibilities of the country. Silver ore, that would assay forty to eighty per cent, in the shape of blue clay, had been trodden under foot, washed away, sluiced out, and gotten rid of in the easiest way possible. It was said there were mountains of it. Previous to this California had had many excitements. Gold Lake, Gold Bluff, Kern River, Frazer River, White Mountain, and others had all drawn away their thousands and sent them back disappointed; but in those instances gold, that was only found in small quantities, was the object sought. The new discoveries were silver ores. Some who visited the new mines reported, on their return, that more millions were in sight at Gold Hill and Virginia than California had yet produced. All the stories of the fabulous wealth that Spain drew from South America and Mexico came to mind; of Spanish galleons sunk with the weight of silver on board; of the solid altars and crucifixes of silver; of the hundreds of vessels with rich cargoes captured by the buccaneers; of cities plundered of their vast wealth; of the burial of the piles of money in many places along the Atlantic and Pacific Coasts, and on lonely islands. The awkward coinage of the Mexican and South American money bore evidence of the rude age, when half-savage miners boiled their frijoles in silver kettles. A new Mexico, a new Peru, was found just over the Sierra Nevada, and the whole country was aroused. As soon as the melting of the snow permitted, and even before, a great multitude set out for the silver land, some on foot, and some with pack-mules.

Gold Hill City, 1860s

Library of Congress, Lawrence & Houseworth Collection

Immense freight-wagons, with ponderous wheels and axles, heavily laboring under prodigious loads of ore for the mills, or groaning with piles of merchandise in boxes, bales, bags, and crates, block the narrow streets. Powerful teams of horses, mules, or oxen, numbering from eight to sixteen animals to each wagon, make frantic efforts to drag these land schooners over the ruts, and up the sudden rises, or through the sinks of this rut-smitten, ever-rising, ever-sinking city. A pitiable sight it is to see them! Smoking hot, reeking with sweat, dripping with liquified dust, they pull, jerk, groan, fall back. and dash forward, tumble down, kick, plunge, and bite; then buckle to it again, under the gaffing lash; and so live and so struggle these poor beasts, for their pittance of barley and hay, till they drop down dead.

J. Ross Browne, 1871

Freight from Gold Hill, Virginia and Truckee Railroad depot

July 13. [1871] George Kirk was hung by Vigilants, at Virginia City. He had been ordered to leave town, and came back; was found drunk in a dance-house, taken to the Sierra Nevada works and hung; had "601" pinned to him.

SUPPLIES FROM CALIFORNIA. The whole of western Utah, or Nevada as it was afterwards called, did not produce provisions enough to supply the new population a week; but California had now become an exporter, and in a short time the roads leading to "Washoe" were thronged with teams carrying everything over the mountains, from quartz machinery down to strawberries, that could be desired.

Ten years of cultivation had developed the agricultural resources of California, and the miners of the new Territory could make themselves far more comfortable than did the gold miners in the days of '49. A passable wagon route across the mountains, used by the first emigration, enabled the farmers of El Dorado and the adjoining counties to carry in provisions, but soon costly roads were established, with easy grades, which were kept sprinkled, and equal to the walks of a city.

TEAMSTERS' ASSOCIATION. Thousands rushed into teaming, but it was by no means a sinecure, though there was enough profit to induce hundreds of men to engage in it. Freight at first was enormously high, twenty-five cents a pound not being deemed too much for hauling over the rough roads. Finely graded roads enabled the teamsters to make money at two cents a pound, or forty dollars a ton, and the competition became so sharp that a "Teamsters' Association" was established, which fixed the uniform rate at sixty dollars per ton from Folsom, the terminus of the Sacramento Valley Railroad.

It became well understood that goods shipped through other agencies were liable to be injured while in transit. Sometimes the wagon containing them would unaccountably roll over the grades in a dark night, while the owners were camped but a few feet away. Again, nuts from the wheels would be missing; harness would be cut, and a man known to be "cutting under" was annoyed in various ways.

ROAD AGENTS. As highwaymen designated themselves, drove a thriving trade during the early days of the Washoe excitement. Provided they escaped the first wrath of the victims the robbers were generally safe enough, for few persons had any time to track up a thief, or prosecute the case in court. As no one thought of traveling without money, almost everyone, even the man trudging along on foot, would have fifteen or twenty dollars, and a few days of successful foraging in this way would make quite a "stake" for a gambler or broken prospector. The vacant ground between Virginia City and Gold Hill, as also down the road towards Dayton, was a favorite ground for robbing footmen. Many a man has been halted in a dark evening and compelled to give up his loose change, and many a man who resisted has been shot and unceremoniously tumbled into

International Hotel, Virginia City, 1860s. The International Hotel was six stories high and had the West's first elevator, called the "rising room."

June 10, 1868. The overland stage was met by three men with double-barreled shot guns, and the passengers—four gentlemen and two ladies—ordered out. The ladies were not molested, although one of them had $900 on her person, but the men were relieved of about $4,000.

Cheap Cash Store, Frank Folsom, owner, Gold Hill, Nevada, 1890

October 31, 1866. The stage was stopped on the Geiger Grade, and the safe, containing $5,150, was taken and blown open. The passengers also lost several thousand. Wells, Fargo & Company offered $9,000 for the apprehension of the robbers.

some of the numerous abandoned shafts which dotted that part of the country. Others, bolder in their operations, would attack the stage and capture the bullion which was sent over the mountains in bars.

CAPT. EDWARD FARIS STOREY. After whom Storey County was named, was born in Jackson County, Georgia, July 1, 1828, his father being Col. John Storey, who was in command of a regiment of volunteers during the difficulties with the Indians in the western part of Georgia during General Jackson's term of the Presidency, and afterwards acted as commander of an escort which conveyed them to the Indian Territory at the final settlement of the difficulty. He raised a company of riflemen, and with others made the attack on the fortified camp of the Pah-Utes June 2, 1860, which resulted in the defeat of the Indians. Captain Storey here met his death at the hands of an Indian who, ambushed behind a rock, shot him through the lungs, producing death the same evening.

POLITICAL EXCITEMENT. In Gold Hill the election of Trustees, under the late Act of incorporation, occurred June 6, 1864, and resulted in the success of the Citizens Ticket by 186 majority. C. S. Coover, S. H. Robinson, H. O. Blanchard, Moses Korn, and G. W. Aylsworth were the successful candidates. Great excitement prevailed. Twenty-one double votes, found in the ballot-box, were rejected. One side charged fraud, while the other claimed the result as the "triumph of law and order."

Courtesy of the Nevada Historical Society

AMUSEMENTS IN EARLY DAYS.

These partook of the character of the people; something strong for miners; no milk and water exhibitions. If it had been possible to hang men and afterwards resuscitate them or blow them from cannons and afterwards gather the scattered fragments together it would have drawn finely.

Prize fights were not uncommon. In consequence of being prohibited by law they were generally held in out-of-the-way places. They commenced as early as 1863. [On] January 8,1864. Two Hibernians, "jist to honor Jineral Jackson " arranged a fist fight to come off on a vacant lot on B Street, but the police interfered with the amusement, to the disgust of many of the spectators.

June 4th, Bill Davis and Patsey Dayley fought at American Flat for $1,000 a side. Three thousand spectators witnessed the exhibition at $2.50 a head.

October 1, 1865. A bear and dogfight came off at the Opera House, in which the bear made short work of whipping the dog. After the dog was whipped the police arrested the managers.

GREATER PROSPERITY INDICATED. "Wild cat" schemes are pressed to the front. This term has been used for half a century or more to denote baseless projects. The silver mines of Nevada had more wild cats to the square mile than any other land ever discovered. If men in other places were bitten by them, here men were devoured, lost, so that not a vestige was left.

[1861] McKenzie was killed by Sam Brown, in Virginia City. Brown ran a knife into his victim, and then turned it around, completely cutting the heart out, then wiped his bloody knife and laid down on a billiard table and went to sleep.

Gold Hill from Belcher Dump.

Steve Crandell Collection

...we took our places on the stages, and girded up our loins for the trip across the mountains. I was the lucky recipient of an outside seat. The driver was Charlie. Of course every body knows Charlie—that same Old Charlie who has driven all over the roads in California, and never capsized any body but himself. On that occasion he broke several of his ribs, or as he expressed it to me, "Bust his sides in." I was proud and happy to sit by the side of Charlie. Possibly I may have travelled over worse roads than the first ten miles out of Placerville . . . there are not many quite so bad on the continent of North America. I speak of what the road was at the close of summer, cut up by heavy teams, a foot deep with dust, and abounding in holes and pitfalls big enough to swallow a thousand stages and six thousand horses without inconvenience to itself. There are places, over which we passed after dark...where the horses seem to be eternally plunging over precipices and the stage following them with a crashing noise, horribly suggestive of cracked skulls and broken bones. But I had implicit confidence in Old Charlie. The way he handled the reins and peered through the clouds of dust and volumes of darkness, and saw trees and stumps and boulders of rock and horses' ears, when I could scarcely see my own hand before me, was a miracle of stage-driving. "Git aeoup!" was the warning cry of this old stager. "Git alang, my beauties!" was the natural outpouring of the poetry that filled his capacious soul

J. Ross Browne, 1871

Fredrick House, D Street, Virginia City

Mines were incorporated on ground that did not have a particle of mineral, this being supplied from other mines. Gold-dust was shot into the ground, silver was melted into the rock or plugged in, in such a way as to resemble natural ores, so that a person not an expert would see silver all around in a worthless mine.

Some amusing things in this connection occurred in an early day. A party from San Francisco who had been visiting the mines, returning with a sack of ores stopped all night at a hotel in Amador County with a notorious wag by the name of Hosley. After listening to their talk awhile he planned a surprise for them, and, after they had retired, judiciously exchanged their worthless specimens for similar looking ones, which he knew to be good. The unsuspecting travelers continued on their way to the city where they put their find in the hands of an assayer. The results exceeded their most sanguine hopes. Companies were formed and money raised to carry on the work, but the clouds, though bright, had no silver lining.

Leading Industrial Enterprises.

The gas works were established in November, 1863.

The Virginia and Gold Hill Water Company was the nucleus of the company which afterwards merged in the present Company.

The California Stage Company ran a daily line connecting with the approaching Central Pacific Railroad at Auburn, and also running to Marysville, Grass Valley, Nevada, and other places, also connecting with their line to Portland, 710 miles distant.

Pioneer Stage Company ran a daily line to Sacramento via Gold Hill, Silver City, Carson City, Genoa and Placerville, carrying Wells, Fargo & Co.'s express and the United States mail.

Pacific Express and Stage Company ran a daily line to Sacramento through the Henness Pass, connecting with the steamer at Sacramento, or connecting at Newcastle with the Central Pacific Railroad.

Moving ore carts through the tunnel of the Gould & Curry Mine in Virginia City, 1860s

Library of Congress, Lawrence & Houseworth Collection

Over the Sierra the wire was attached to the trees, and their swaying by the wind, caused the wire to stretch, until, in many places, it lay along the ground between the points of support. It is said that teamsters would sometimes cut out pieces of the line and use it in repairing the wheels of their wagons. One teamster being remonstrated with for this, said he supposed the wire had been placed there by the Toll-road Company to be used for that purpose. In consequence of these breaks, messages were often delayed. If there were important messages passing through and the line was broken the message would be transferred to the Pony Express, and in this way the telegraph was often beaten into Sacramento by the pony rider. This was the case with President Lincoln's first message and the news of his first election.

Miners in Austin in the 1870s

We stopped for a short time at Carson City, at one A.M. Here we got out and stamped around for a few minutes while the horses were being changed, and were amused by a lady who had no money wherewith, to pay her fare any farther, and at the same time declined to alight. The mail agent was in an awkward fix: he did not like to engage in a fray in the dead hours of the night, as the awakened neighbours would be sure to side with the woman they did not know, for the pleasure of abusing the man they did know; and yet if he allowed her to proceed, the amount of her fare would be charged against his pay. At last, however, he was persuaded to leave her in possession by her assurance that she was a person of great consideration, owning houses and lands in Virginia City, and that everybody knew where she lived. So I poked my head into my air-pillow again and off we went.

Edmund Hope Verney, 1865

The Overland Stage Company left daily, westward for Sacramento, and eastward for the Missouri River, passing through Austin and Salt Lake.

They ran full at high rates and consequently could afford to stock their roads with the best of horses. The Fulton Foundry was started in 1863, and in 1865 were ready to make castings or machinery of any size.

The Gould & Curry Foundry did their own work exclusively.

Beer was not forgotten, and five breweries could scarce supply the people with beer, for Nevada has a dry climate!

OVERLAND MAIL STAGE COMPANY. The year following the establishment of the Pony Express, the Southern Daily Overland Mail, which had been established in 1859 through northern Texas to California was transferred to the Central or Simpson route, its regular trips commencing on the first of July, 1861. The reason of this transfer was the anticipated disturbances along the southern line, consequent upon the war of the Rebellion. The trans-continental telegraph was also built along this line. The work of constructing it was commenced in 1859, pushed rapidly forward in 1860 and 1861, and completed the twenty-second of September of the latter year. Previous to the establishment of the whole line, that portion between Placerville and Virginia City was built and operated by the "Placerville and Humboldt Telegraph Company," and was known as "Bee's Grapevine Line," having been projected and built by Col. F. A. Bee.

THE NEWSPAPER DEPARTMENT. The pioneers of Nevada were eminently a reading people. They might plead guilty to charges of extravagance, excitability and recklessness, but no one ever suspected them of a want of general intelligence. The newspaper followed closely the saloon, and when the matutinal drink was taken the morning paper was read as a matter of course.

It will be seen that Virginia City was second only to San Francisco for the number and ability of its papers. The circumstances under which the city had its birth and growth, the class of readers unusually intelligent and energetic, with the large admixture of the reckless and even criminal element in the population, called for editorial ability of the highest class. Firmness, mingled with discretion, honesty without bigotry, and

Sacking the tailings at the Gould & Curry Mine, Virginia City, Nevada

Library of Congress, Lawrence & Houseworth Collection

I moralize well, but I did not always practise well when I was a city editor; I let fancy get the upper hand of fact too often when there was a dearth of news. I can never forget my first day's experience as a reporter. I wandered about town questioning everybody, boring everybody, and finding out that nobody knew anything. At the end of five hours my notebook was still barren. I spoke to Mr. Goodman. He said: "Dan used to make a good thing out of the hay wagons in a dry time when there were no fires or inquests. Are there no hay wagons in from the Truckee? If there are, you might speak of the renewed activity and all that sort of thing, in the hay business, you know.

It isn't sensational or exciting, but it fills up and looks business like."

I canvassed the city again and found one wretched old hay truck dragging in from the country. But I made affluent use of it. I multiplied it by sixteen, brought it into town from sixteen different directions, made sixteen separate items out of it, and got up such another sweat about hay as Virginia City had never seen in the world before.

Mark Twain, 1872

Two nonpareil columns had to be filled, and I was getting along. Presently, when things began to look dismal again, a desperado killed a man in a saloon and joy returned once more. I never was so glad over any mere trifle before in my life. I said to the murderer:

"Sir, you are a stranger to me, but you have done me a kindness this day which I can never forget. If whole years of gratitude can be to you any slight compensation, they shall be yours. I was in trouble and you have relieved me nobly and at a time when all seemed dark and drear. Count me your friend from this time forth, for I am not a man to forget a favor."

If I did not really say that to him I at least felt a sort of itching desire to do it. I wrote up the murder with a hungry attention to details, and when it was finished experienced but one regret--namely, that they had not hanged my benefactor on the spot, so that I could work him up too.

Mark Twain, 1872

Shaft #1, Sutro Tunnel, 1860s

the ability to treat with vigor all the current questions of the day, were absolute essentials without which a paper would not survive a week.

PRINCIPAL FIRES IN VIRGINIA CITY.

The first great fire in Virginia City broke out August 29, 1863, in a carpenter shop in the rear of Patrick Lynch's saloon. About $700,000 worth of property was destroyed.

On September 29, 1865, a fire started at the Fountain Head Restaurant. It burned over an area extending from Union Street to below Sutton Avenue, and as far as D Street east, and A Street west. About $400,000 worth of property was destroyed.

On September 23, 1866, Music Hall was destroyed by a fire occasioned by the bursting of a lamp.

June 29, 1873, at 11 o'clock, p. m., the McLaughlin & Root building, on B Street, blew up and took fire; 100 pounds of Hercules powder, six cases of nitroglycerine, 100 pounds of giant powder, and 200 pounds of common powder had been stored under the bed-room of Major General Van Bokkelen, by that gentleman, and exploded. He was killed.

The great fire, one long to be remembered, commenced at 5:30, A. M., October 26, 1875, in a low lodging-house on A Street, and resulted in the total destruction of the business part of the city, and a loss of about $12,000,000. The "fire-proof" buildings seemed to offer as little resistance as those of wood. The mills and hoisting-works were swept away as by a whirlwind.

The shafts of the mines burned down to a considerable distance, occasioning much caving in. At the Ophir a cage was let down and covered with dirt to prevent the fire from passing down, but the fastenings or springs gave way when the dirt was shoveled on it, and the fire went down the shaft.

The people set to work to rebuild even while the beds of coals were glowing with heat, and in a few days most of the people were under shelter. The railroad brought in supplies of timber and provisions. Forty-six trains passed over the road in one day; 100 cars were dropped at Reno for Virginia City in one day". But for the railroad the city must have been abandoned until spring. Let those who see no good in railroads make a note.

The mines were soon in working condition. The Ophir shaft was repaired (retimbered) sixty feet deep, new and powerful hoisting works set up, and everything in running order in thirty days, four of which were used in putting out the fire. Samuel Curtis was the Captain in this rapid work.

Virginia had grown to be the "livest" town, for its age and population, that America had ever produced. The sidewalks swarmed with people--to such an extent, indeed, that it was generally no easy matter to stem the human tide. The streets themselves were just as crowded with quartz wagons, freight teams and other vehicles. The procession was endless. So great was the pack, that buggies frequently had to wait half an hour for an opportunity to cross the principal street.

Mark Twain, 1872

The fire chief and perhaps his crew for an 1890's Gold Hill fire company

Steve Crandell Collection

Lemuel S. Bowers, commonly known as "Sandy," was an ignorant, easy-going frontiersman, happening, in 1859, to be mining for gold in Gold Cañon by the simple process of washing the mineral-bearing earth in a rocker, and as developments continued found that his claim of ten feet covered a portion of the Comstock Lode. Adjoining was a claim of the same dimensions belonging to Mrs. Cowan, who also resided in the cañon and was washing and cooking for the miners. The two married, and the claims became one, proving of extraordinary richness. In a few years they were overwhelmed with wealth. Too ignorant of business, they knew nothing of prudent or cautious investments, and became the tools of harpies. The now wealthy couple were advised—as a good joke—to take a tour through Europe to see the sights and become polished in accordance with the station they were in the future to occupy. They were also advised to build a palace worthy such a party to reside in. Accordingly in 1861 the "Bowers Mansion" was commenced in the wilderness of Washoe Valley. Before leaving for Europe Sandy was told that the proper thing to do was to give a banquet. "Banquet goes," said Sandy, and the International Hotel of Virginia City was engaged for the occasion. Every obtainable luxury was ordered which Virginia or San Francisco could furnish. Champagne was to be as free as water in a spring flood. Everybody was invited. Toasts were drank and in response to "Our host," Mr. Bowers was called upon to reply. He arose and delivered the following characteristic speech.

continued in far right column

Consolidated Virginia Mine, Virginia City

[A COLORFUL CHARACTER.] As the story goes, it was in 1850 that Jones (Hon. John P.) and his partners repaired to Sonora on a Sunday, according to the custom of the country, to lay in supplies for the coming week, see the sights, and spend such few scads in pleasures as the state of their purses would warrant. While there a great outcry occurred in one of the corrals. A little, lean, insignificant looking jack, upon which a Mexican was packing his crowbar, batea and frijoles, had wandered unobserved in the corral, and had, notwithstanding his burden, successively attacked and whipped out all the horses in the yard, finishing up with a famous black stallion, whose fighting qualities were beyond question, he being considered not only dangerous, but invincible. But the extraordinary development of jaw in the jackass, combined with a phenomenal courage, enabled him to make short work of the big stallion, that was now writhing, utterly helpless in the terrible mouth of the infernal little animal, that was as relentless as a bear trap. After some considerable trouble the jack was induced to let go his hold, and was led out to be shot; but the Mexican pleading his poverty and the generally peaceable character of the animal, the sentence was commuted to banishment, on condition that he should be instantly taken out of town.

Jones, who had quietly witnessed the proceedings, was struck with an idea. In the camp where he was mining lived an odd character from Arkansas, by the name of Joggles, who owned a worthless old plug of a horse, which had made itself famous and finally infamous by running everything off the range in the shape of a horse. When appealed to in regard to the ill-doings of the brute the old man would laugh until the tears would run down his cheeks, remarking that "Old Pison is some, you bet," and his valuation of the animal would go up with every fresh complaint, until half the money in the camp would not buy him; in fact, the horse had become an intolerable nuisance, but old Joggles was on the shoot and it was dangerous to molest Old Pison. In Jones' opinion the jack was good for him, and a bargain was soon struck with the Mexican, who was glad to get two ounces for the jack. Jones and his partners packed their supplies on the brute, that had by this time resumed his sleepy, innocent look, and about sundown they reached their camp with their purchase. As expected, old Joggles made his appearance, and joined with the crowd in the funny remarks about the new animal.

Office at the mouth of the Sutro Tunnel

continued from far left column

"I've been in this yer country amongst the fust that come here. I've had powerful good luck, and I've got money to throw at the birds. Thar ain't no chance for a gentleman to spend his coin in this country, and thar ain't nothin' much to see, so me and Mrs. Bowers is agoin' to Yoorop to take in the sights. One of the great men of this country was in this region a while back. That was Horace Greeley. I saw him and he didn't look like no great shakes. Outside of him the only great men I've seen in this country is Governor Nye and Old Winnemucca. Now me and Mrs. Bowers is goin' to Yoorop to see the Queen of England and the other great men of them countries, and I hope you'll all jine in and drink Mrs. Bowers' health. Thars plenty of champagne, and money ain't no object." Sandy and his wife spent several years abroad, purchased much elegant furniture, laces and pictures for his mansion in Washoe, which was erected at a cost of over $400,000, and returned, and still had "money to throw at the birds;" the hawks and vultures, and other birds of prey getting the greater portion. Without any good missionary to instruct, or any strong friend to advise and direct he continued to throw money at the birds with the approval and encouragement of flatters, sycophants and robbers, and his princely fortune was wasted. His widow earns a precarious livelihood near the scenes of her former toils— and glory.

The discovery of silver and the development of the mines at Virginia City, gave rise to a rapid increase of trade, and other and competing lines of stages were started. Quick trips from Virginia City were often required to be made by parties on special business to Sacramento, and they were sometimes made in an incredibly short time. On the twentieth of February, 1864, the Pioneer line is reported to have made the trip in five minutes less than twenty-four hours. The fastest time recorded was on Jane 20, 1864, when the Larue line is reported to have made the trip over the mountains, from Virginia City to Sacramento, in twelve hours and twenty-three minutes, carrying the mail and William M. Lent, John Skae, and S. Cook, as passengers, they having chartered the coach.

Potosi Mine, Virginia City

"What ur yer gonter do with thet thar critter ?" says Joggles, referring to the jack.

"Turn him out to grass," says Jones.

"He, he," chuckled Joggles, "he won't be a mouthful for Old Pison; he'll chaw him inter a dish-rag quicker'n shucks."

"Don't know about that," says Jones, "you can't tell how these jacks will fight."

"Bet yer fifty dollars Old Pison licks him in a minute," says Joggles.

Jones— "Well, old man, I don't know as I want your money, but if you want to bet a little on it perhaps we can accommodate you; but then a jack is a mighty ugly thing to fight; you had better not try it."

The old man's blood was up. Old Pison never had been whipped, and his opinion was that it was impossible. "I'll bet yer one hundred," says he; "I'll double the bet." Jones didn't care much to bet, but thought that the jack could whip Old Pison. Joggles continued by raising the bet to $300, and daring Jones to take it. Jones reluctantly consented, provided the jack could rest until to-morrow, which, as it was now nearly dark, the old man agreed to.

Joggles dug up his oyster can containing his pile, and put up the dust. It was agreed that the jack was to be turned out on the flat back of the town, and that they should not be driven together, but left to meet each other "sorter by chance." Jones and his company went to work in sight of the expected arena. After a time Old Joggles was seen driving the horse over the hill from the opposite side, though he kept out of view as much as possible. The two creatures, from an inborn sense of true chivalry, mutually recognized each other as worthy foes, and gave the challenge for mortal combat, the horse, by laying back his ears, elevating his head, and giving a loud snort; the jack, by a series of sharp though graceful curves with his spike tail, and a loud blast from his war trumpet.

The horse rushed to the onset with mouth open, wide enough to take in any part of the jack but his head. The ground on which the meeting occurred was a kind of red clay, and the dust obscured the combatants from view for a few minutes, but when they did come to sight the horse was making for town for dear life, with the infernal jack hanging to his withers. The hold broke loose, and Old Pison put in his best licks, getting away from the jack, who came after as fast as his short legs could carry him, his tail rapidly making short circles in the air, and his terrible trumpet uttering the fiercest notes in his repertoire. Down the hill came the horse, his eyes standing out as if pursued by a fiend. The fight was all gone out of him now. In abject terror he rushed to his stable for security, but the door was closed, and Old Joggles was some distance away, following

US Geological Survey office in Virginia City, 1870

Denver Public Library, Western History Collection

I discovered some emigrant wagons going into camp on the plaza and found that they had lately come through the hostile Indian country and had fared rather roughly. I made the best of the item that the circumstances permitted, and felt that if I were not confined within rigid limits by the presence of the reporters of the other papers I could add particulars that would make the article much more interesting. However, I found one wagon that was going on to California, and made some judicious inquiries of the proprietor. When I learned, through his short and surly answers to my cross-questioning, that he was certainly going on and would not be in the city next day to make trouble, I got ahead of the other papers, for I took down his list of names and added his party to the killed and wounded. Having more scope here, I put this wagon through an Indian fight that to this day has no parallel in history.

My two columns were filled. When I read them over in the morning I felt that I had found my legitimate occupation at last. I reasoned within myself that news, and stirring news, too, was what a paper needed, and I felt that I was peculiarly endowed with the ability to furnish it.

Mark Twain, 1872

Inside the Consolidated Virginia Pan Mill, Storey County

December 10, [1862] Jack Williams, a noted desperado, was killed in Pat Lynch's saloon. Pistols were fired in the front room to attract attention, when the rear door of the back room was opened a few inches and a shot fired from a pistol, which killed him. He had killed several men in California and Nevada, had bitter enemies, and expected to be killed finally. He was out on bail for robbery at the time.

up the fight. Old Pison paused just a moment, but the jack was coming, with that terrible mouth distended, for another bite, and as there was no time to consider the situation, he sorrowfully passed on through the town; but the road terminated in a deep gully over which it was impossible to leap, and into which it was death to jump.

Old Pison paused a moment on the brink, but the enemy was upon him; over he went, choosing death rather than another encounter with that terrible pair of jaws. When Joggles got on the ground he beheld his favorite just expiring, and the jack looking on, venting his still unsatisfied rage in furious trumpetings.

"Dog on yer big coffin head, yer've licked Old Pison. Nothin' can't live that's did that;" and, drawing his revolver, the jack was soon lying in death with his defeated enemy.

"Jones," says the old man, his revolver still in his band, "you know'd how that there crittur cud fight." Jones had need of all his diplomacy to make the old man believe that he didn't know; but peace was made, and Old Pison never troubled the range again.

The Bonanza Period.

If the discovery of the silver mines had startled the commercial world, the bonanzas had the effect to astonish and move it to an incredible activity. As it became known that greater deposits below the surface had been found than were ever known before, when millions on millions began to roll into the banks and mints, it had much the effect on trade and commerce of the first knowledge of the abundance of gold in California. Where would the new adjustment of values cease? Whatever else might betide, a dollar would remain a dollar, though it might not purchase as much food. Clothing, houses or lands as in former times it would still pay a dollar of indebtedness. Up to 1865 the yield of the Comstock Lode had been about $45,000,000.

This section of time, although named the Bonanza Period, opened with a few gigantic financial operations, each of which in any other part of the world or at any other time, would have been considered as brilliant, daring or reckless, as risky or safe

Piper's Opera Hall, Virginia City

Courtesy of the Bancroft Library, University of California, Berkeley

". . . the saloons along the board sidewalks are glittering with their gaudy bars and fancy glasses, and many-colored liquors, and thirsty men are swilling burning poison: organ grinders are grinding their organs and torturing their consumptive monkeys; hurdy-gurdy girls are singing bacchanalian songs in bacchanalian dens . . . All is life, excitement, avarice, lust, devilry, and enterprise."

J. Ross Browne, 1863

Mouth of the Sutro Tunnel, commenced October 19, 1869. Photo 1870s

[Traveling with Adolph Sutro] Eight left Virginia yesterday and came down to Dayton with Mr. Sutro. We found Dayton the same old place but taking up a good deal more room than it did the last time I saw it.

We trotted briskly across Ball Robert's bridge. I remarked that Ball Robert's bridge was a good one and a credit to that bald gentleman. I said it in a fine burst of humor and more on account of the joke than anything else, but Sutro is insensible to the more delicate touches of American wit, and the effort was entirely lost on him. I don't think Sutro minds a joke of mild character any more than a dead man would. However, I repeated it once or twice without producing any visible effect, and finally derived what comfort I could by laughing at it myself.

Mark Twain, *Territorial Enterprise* [excerpt], November 1863-February 1864

principles dominated the observer. Among the most prominent measures inaugurated and under way about this time may be mentioned the Sutro Tunnel; the works for bringing water to the Comstock from the Sierra Nevada, and the Virginia and Truckee Railroad. Each of these became, in the hands of the projectors, a great factor in the tremendous drama which the money gods of the Pacific Coast acted during the years of the discovery of the bonanzas.

THE SUTRO TUNNEL. This project, though favored at first by all the mining companies, came to be regarded finally, in consequence of local interests which it affected, as the bête noir, the death's head and cross-bones of every vested interest in the county. The projector, Adolph Sutro, however, proved himself no mean competitor with any who entered the drama, fighting his way inch by inch, and stubbornly holding every, coigne of vantage, whether among the miners of the Comstock, in the Legislature of Nevada, in the Halls of Congress, or among the capitalists of Europe.

THE VIRGINIA AND TRUCKEE RAILROAD. Was also a daring enterprise. The country to be supplied by it was of limited extent, the route which it was to traverse, a mountain region of precipitous cliffs and deep gorges over which it was deemed a triumph of engineering to carry a wagon road. In any country and by any other people the project would have been deemed chimerical, but it was carried through, and became a powerful operator in the period under consideration.

On the seventh of November, 1871, the road was completed from Reno to Steamboat Springs, and the first train passed over the road between those points. On the twenty-fourth of the following August, the last spike was driven that completed the line from Reno to Carson on which day the first train passed over the road from Virginia City to the Truckee River. The first freight from Reno to Carson, all the way by rail, was the press and material for use in the new Appeal office. On the nineteenth of September, the first through freight cars, two of them from San Francisco to Virginia City, passed over the road. On the first of October, 1872, the first regular passenger train passed over the line, with Harry Shrieves as conductor. In December, 1872, the company commenced the construction of their car and machine shops in Carson, and on the eleventh of the same month, the construction of the telegraph line from Reno to Virginia City along the railroad was commenced. In 1874 steel rails were laid between Carson and Virginia, necessitated because of the large amount of business, thirty-six trains per day being required to carry the passengers and freight.

WELLS, FARGO & CO'S EXPRESS. There was an institution peculiar to the Pacific Coast, which has no such "office" but stood ever ready on the frontier, and wherever the miner pitched his tent, however broad the desert or rugged the cañon, if letters were to be sent or bullion carried, there went the messenger with his pouch and strong box. This institution was Wells, Fargo & Co.'s express, always in the van of pioneers, ready with

"The Comstock Lode has up to this time been traced more than twenty thousand feet, or nearly four miles. It runs nearly due north and south, and is popularly divided into three portions, the northern, the middle, and the southern; or, as others style it, the Ophir, the Virginia, and the Gold Hill. On these two latter portions stand the towns of Virginia and Gold Hill; literally stand on them, for the tunnels and drifts run under the towns, sometimes with so thin a separation that, as happened in Virginia City, two or three houses and a church paid a visit to the depths of the Gould and Curry mine, or at least went part of the way down."

J.G. Player-Frowd, 1871

Ore train crossing the Union Trestle in Virginia City

One ton of bullion ready to be shipped from the Wells Fargo & Company Express office in Virginia City, probably between 1880 and 1900

July 18. [1868] Peter Hill, alias "Russian Pete," while resisting arrest for robbery at Silver City, Storey County, took refuge in the North Potosi Tunnel.

While the officers and posse were attempting to drown him out, he killed one of the posse, and then putting the pistol into his own mouth blew his brains out.

the rush to go, serving its purpose and reaping its reward, then retiring as business declined, its facilities and accommodation always corresponding with the times. This company rendered the pioneers needed service, for which it is held in grateful remembrance. So prompt and faithful were its messengers in the delivery of letters, that for several years the express did the principal carrying business, charging but two to seven cents in addition to the United States postage. In addition the company transported all the bullion of the country, keeping such a record of its production that its statistics have become authority superseding all others.

THE WATER SUPPLY. It was not until 1873 that the inhabitants of Virginia City and Gold Hill enjoyed an abundance of pure, soft water. In the early days natural springs afforded a sufficient supply for the few persons living in the two mining camps. As the population increased these springs were found inadequate to meet the demands of the people, and various devices were adopted to collect and distribute the water flowing from several tunnels which had been run into the mountain west of Virginia City for prospecting purposes. Large wooden tanks were built at different points to store the precious fluid, but the company which had been organized to supply the community frequently found itself embarrassed in its attempts to keep filled these rude reservoirs. The tunnels running dry, a water famine would be imminent; when new strata of rock were cut across and for a time the supply increased. But the tunnels at the best furnished but feeble streams, and these were charged with minerals. The next device resorted to was to dam up the shallow basins on the summits of the distant hills to hold back the water from the melting snow. These were found to yield largely and for a long time, when tapped by a tunnel run under the basin or sunk at the depth of 300 or 400 feet. Yet one after another these hills failed. Thousands of dollars had been expended in these various experiments, but the danger of water famines constantly confronted the people. Finally the Virginia and Gold Hill Water Company determined to bring a supply of pure water from the streams and lakes of the Sierra Nevada Mountains—from the regions of eternal snow, It was a bold scheme and its accomplishment one of the most remarkable engineering triumphs of the age.

THE FOUNDRIES OF VIRGINIA CITY. The mines of Storey County, with that extensive demand for castings and machinery of all kinds, led to an extraordinary development of the foundry business. The great foundries of San Francisco are indebted mainly for their rise and prosperity to the discovery of silver on the Comstock. At an early day, however, it was seen that an inviting field for the foundryman's industry was to be found in the vicinity of the mines. The pioneers in Nevada in this branch of industry were Messrs. Mead, McCone & Tascar. Those gentlemen had for a long time conducted a flourishing little foundry in Placerville, California. In the fall of 1862 they

"I have a suggestion to make. That is, that the Government employ, at a living salary, George Francis Train and the Clafflin family, to create a current which shall knock the earth off its orbit, cause it to turn a summersault and reverse the poles, and trust to a kind providence to restore order and the desired equality of climate. Without doubt Virginia City could furnish the giant powder and nitro-glycerine, and the subterraneous caverns where the explosives could be planted. The melted lava needed for mixing comets could be brought from South America in a balloon, as it is not a safe commodity to put on board vessels. This advice I give partly because I feel an affection for a people who need more moisture, and partly from the fact that I, as an individual, would like to establish a reputation for being a philosopher as well as a political economist."

Caroline M. Nichols Churchill, 1874

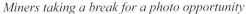

Miners taking a break for a photo opportunity

Courtesy of the Library of Congress

J. Jones, Jr. Pharmacy in Gold Hill, 1875

Here were some thousands of excited men, accustomed to the use of fire-arms from infancy, who had invested largely in the Love's Delight, Sorrowful Countenance, Pious Wretch, Literary Cuss, and other valuable claims of a kindred character—all awaiting, with stern resolution and ill-suppressed rage, the coming of this diabolical quill-driver, who had so basely ruined their mines and blasted all their prospects. Many thousands of people had no other idea of Washoe than what they gathered from these ridiculous caricatures, which were a monstrous fabrication from beginning to end.

J. Ross Browne, 1871

moved their machinery over the Sierra, and established themselves in Johntown, two miles below Silver City.

In 1869 Greeley sold the Nevada Foundry to the Bank of California, when the latter placed it in charge of Mr. Graves, a master-mechanic on the Central Pacific Railroad, and Mr. J. M. Quimby, also a railroad man. All the castings and finishings for the Virginia and Truckee Railroad were manufactured by the establishment at this time.

THE MANUFACTURE OF ICE. The water company put up an ice factory in 1877 using Holden's Machine, that is capable of making fifteen tons of ice daily, which is sold for about twenty dollars per ton. Ice was formerly brought from Truckee.

SOME OF THE LEADING MINES

The Ophir Mine is one of the oldest, if not the oldest discovery of the group, having been mined since early in 1859, the bonanza reaching to the surface. It has yielded over $10,000,000 in bullion, and declared dividends to the amount of $1,504,400. The assessments have been $3,088,200.

Best & Belcher is another of the promising mines, the assessments reaching nearly $500,000. Ore bodies have been reported as existing, but no product of bullion has proven their value. The mine consists of 540 feet on the lode, and is one of the oldest locations.

The Gould & Curry Silver Mining Company was incorporated on the twenty-seventh of June, 1860.

The claim of the company is centrally located on the Comstock Lode, and has yielded $15,644,220.63 in bullion; most of which has been extracted from above the adit levels.

A prospecting shaft, inclined below the 1,500-foot level, has been extended to a vertical depth of 1,900 feet, disclosing, so far as explored, a vein of undetermined width of very promising ore-bearing material.

Owing to the great expanse of the lode at this central position, it was judged advisable to suspend the prospecting operations from this incline, and resume its exploitation from, a point nearer its eastern confines.

Another shaft was accordingly commenced 2,285 feet still further to the eastward, which has, at the perpendicular depth of 1,970 feet, or 2,370 feet below the croppings, penetrated the easternmost borders of the ledge formation.

The enormous expenditure incurred in sinking this east shaft, although primarily a severe tax on the shareholders, will ultimately be of great advantage in economically working the mine.

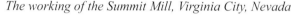

The working of the Summit Mill, Virginia City, Nevada

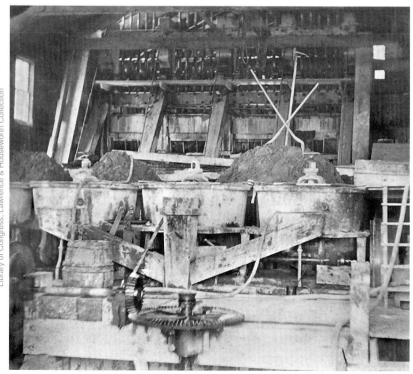

Library of Congress, Lawrence & Houseworth Collection

The "Gould & Curry" company were erecting a monster hundred-stamp mill at a cost that ultimately fell little short of a million dollars. Gould & Curry stock paid heavy dividends--a rare thing, and an experience confined to the dozen or fifteen claims located on the "main lead," the "Comstock." The Superintendent of the Gould & Curry lived, rent free, in a fine house built and furnished by the company. He drove a fine pair of horses which were a present from the company, and his salary was twelve thousand dollars a year.

Mark Twain, 1872

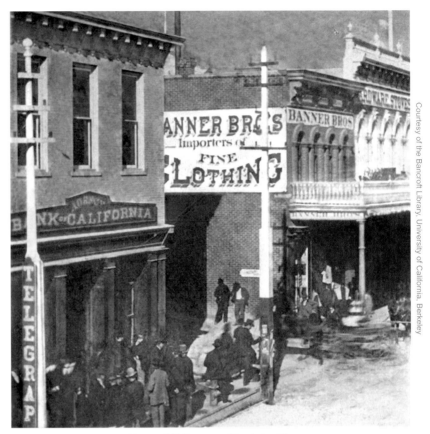

View of C Street in Virginia City

As I grew better acquainted with the business and learned the run of the sources of information I ceased to require the aid of fancy to any large extent, and became able to fill my columns without diverging noticeably from the domain of fact.

Mark Twain, 1872

As the necessary buildings are finished, and their equipment complete, the cost of continuing the shaft to an additional depth of 1,500 feet will be comparatively small, and will be borne in part by the adjacent mine (the Best & Belcher Company).

Such is the magnitude of the mineral lead within the boundaries of this mine, that it is calculated this extreme depth will have been attained before the west wall of the ore formation shall have been reached.

The character of the rock exposed in the deepest workings continues similar to that which inclosed the bonanza existing near the surface.

In the new shaft, as it progresses downward, the indications of the close proximity of another ore body are rapidly increasing.

At the last annual election, held in San Francisco, on the twentieth of December, 1880, the following officers were elected: W. S. Hobart, President; A. K. Durbrow, Secretary; H. H. Penoyer, Superintendent.

It is confidently expected that a development of value will have been made in this mine before the next annual meeting, which will be satisfactory to all the stockholders.

The works of this company are among the finest on the Comstock, and can be better appreciated by turning to the view, to be found on another page of this volume.

The Confidence has had a body of paying ore, and paid $78,000 in dividends, and has also expended something over a quarter of a million in assessments. It was worked through the Yellow Jacket shaft, the ore body being a part of the Yellow Jacket bonanza.

Yellow Jacket Silver Mine -- located in the spring of the year 1859, by Bishop, Camp, Rogers, and others, and consists of 957 feet of the Comstock Lode; It has been worked continuously since its location, and has produced $14,372,172. The company was incorporated February 17, 1863, under the then existing laws of the Territory of Nevada, and has continued a Nevada incorporation, being the only mine on the Comstock Lode that has its home office at the mine or in the State of Nevada. The present number of shares is 120,000, of the par value of $100 each.

April 9. [1868] Charles Watson was killed by George Newton, at Silver City, Storey County. They commenced fighting up-stairs, and rolled down locked in each other's embrace, when the fall broke their holds. Newton got out his pocket-knife and stabbed Watson.

Silver City from the South end of town

Courtesy of the Library of Congress

Unknown mining works, Virginia City

". . .turning sharp round the base of Mount Davidson come to the Southern or Gold Hill mines, the rich, irregular, coquettish, delusive mines of Gold Hill. More speculation in shares and more fortunes have been made and lost in these mines than in any of the others . . . Then came one morning a telegram, 'The Yellow Jacket is on fire,' followed by other telegrams: 'The fire has extended to the Crown Point and Kentuck.' There was great loss of life in the lower levels, as they were cut off by the fire which commenced in the third level.

At the date of the fire (April 7, 1869) Yellow Jacket had 5,000 tons of twenty-seven dollar ore exposed on its nine hundred foot level. Kentuck had 12,000 tons of thirty dollar ore exposed between the seven and nine hundred feet level; but the fire, which a year after was only partially extinct, has destroyed the timber work and caused the caving of much of the partially worked ground between the six and nine hundred feet levels, especially in the Kentuck mine."

J.G. Player-Frowd, 1871

The Kentuck Mine comprises ninety-four feet on the Comstock Lode, next south of the Yellow Jacket, and is one of the locations of 1859. A rich body of ore came to the surface through nearly all these Gold Hill claims, and yielded many millions of dollars before barren ground was reached. The Kentuck was a long time in bonanza, and, up to 1870, had paid $1,252,000 in dividends.

The Mexican Mine is a non-paying mine, valuable for its possibilities. The present company was incorporated in 1874. Total assessments levied, $1,436,000. It derives its prospective value from its vicinity to the Ophir, through the shafts of which it has been explored to a depth of two thousand feet or more.

The Overman Mine is one of the most noted and important in many respects on the Comstock. It is a point of departure for the systems of mines which terminate in American Flat and Dayton, having its east and west ledge, both of which have been considered as promising investments. The west ledge in early days produced a considerable quantity of bullion, enough to induce thorough prospecting. Subsequently a new shaft was sunk, near 1,500 feet east. The assessments have been in the vicinity of $3,000,000. The amount of bullion is estimated at $3,239,400.

Accidents in the Mines.

Mining, by general consent, is conceded to be a dangerous occupation. The utmost care on the part of Superintendents can not avert all danger. The great depth, the eternal darkness, dispelled only by the feeble light of a tallow candle; the giving away and crushing of the timbers in some of the numerous chutes and drifts, precipitating rock or dirt down upon the miner hundreds of feet below; the generation of poisonous or explosive gases; the danger from floods of water, which may come at any time with overwhelming rapidity; and last, but not least, fire, all combine to make deep mining one of the most dangerous avocations which can be followed. It has been said that the deaths from accident in the Comstock mines average one a month. Sometimes there are none for weeks, then they may succeed each other with startling rapidity; but the generality of them have so accustomed themselves to see a man brought out of the shafts maimed, limp, and lifeless, or torn in pieces,

July. [1867] Tucker John, a Pah-Ute, was killed at one of Coffman's stations, on the Humboldt road, by Alexander Fleming, of Dayton. Fleming suspected the Indian of killing his brother some three years since.

Mexican Mine, Virginia City

Courtesy of the Library of Congress, Lawrence & Houseworth Collection

Passenger car for the Sutro Mine

November 9. [1870]
W. G. Snell was killed in a mining dispute over the Banner and Creole mines. Also resulted in wounding ten others and throwing two men down a shaft seventy feet.

that, beyond a passing remark, it excites no comment, being regarded as a thing of course.

THE YELLOW JACKET DISASTER, occurred April 7, 1869, was so fatal in its mortality that the date has been reckoned as the black day. The fire started in the 800-foot level about seven A. M., and was doubtless caused by some one of the retiring night-shift leaving a candle among the dry and almost half-charred timbers which have taken the place of the ores extracted at that depth, A part of the day's shift had been lowered into the Yellow Jacket, Crown Point, and Kentuck before the flames burst out. When the peril was discovered the fire alarm was sounded, and the fire companies of Gold Hill and Virginia City responded with alacrity. Simultaneously with the fire alarm, the smoke, thick and dark, was seen coming up from the shaft, and then it was known through both towns that men were being burned in the mines, or smothered by the noxious gases. Many of the miners who were perishing below had wives and children in the town. These, with others, came to the works. When they saw the hopeless situation they had to be restrained from, throwing themselves into the burning pit, for the instinctive thought of woman is, that "if I were only there I could do something for them." The fire companies could do but little towards staying the fire, and but a few were got out alive, and these by retreating into the adjoining mines. Some were suffocated while flying along the lower galleries; some made their way to the shafts only to fall into the devouring flames. The sulphurous vapors generated by the fierce fire against the mineral rocks filled the lower levels, and rendered it almost impossible to recover the bodies even. During that and the following day twenty-three bodies had been recovered.

...miners, braving death in so many forms, become reckless, seemingly balancing themselves on the brink of destruction, with little care which way they fall. Let one unused to mining stand at the mouth of a deep shaft that goes 2,000 feet or more down into the earth, and see the men scuffling for places on the tub or cage; see them clinging on the outside, where the slightest indiscretion will precipitate them against jagged rocks 1,000 feet below, or subject them to the danger of having an arm or head torn off against the timbers of the shaft while descending in the bucket, and the wonder is that more are not killed.

A party ready to descend Comstock Mine. Men (including McCormick, James Horsburgh, Jr., Professor Joseph N. LeConte, and a Black man) pose with candle lanterns before a visit to the Comstock Mine, in Virginia City, Nevada. Late 1800s.

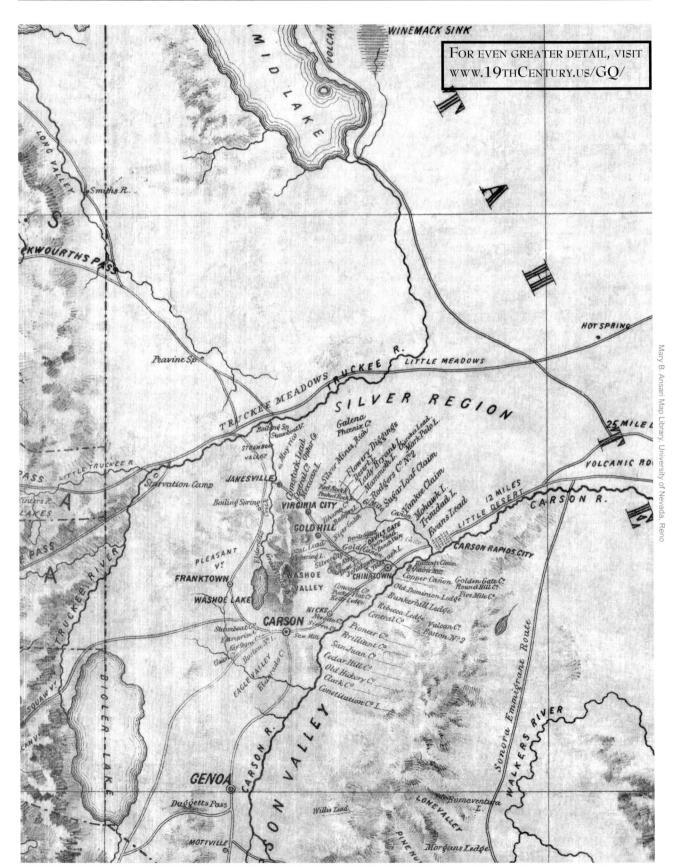

Map of the Washoe Mines, 1860

Freight depot at Reno, 1860s

Chapter 5: Washoe County - Reno and vicinity

Although it is probable that some of the American trappers that penetrated into this region as early as 1832 may have traversed the country now embraced within the limits of this county, yet there is no record of any visit prior to that of Lieutenant Fremont, on his second exploring expedition. He came down from Oregon through Roop County [Nevada], and, on the tenth of January, 1844, discovered and named Pyramid Lake, and on the fifteenth came to the mouth of the Truckee River, which he named Salmon Trout River.

Though not the earliest settled portion of the State, still Washoe Valley was known and Truckee Meadows were known by the earliest emigrants, those that passed through to California

Dear Mary,

...there was no grass for sixty-five miles and but one spring, a mile off the road, where water could be had for the cattle; in short, we were on the desert and drove the whole distance without feeding our cattle, and no water except at the commencement. Our train was the fourth that had taken the road, and I counted on the last thirty miles fifty oxen dead from exhaustion on the desert.

Alonzo Delano, 1849

Dear Mary,

...A man deserves to be well paid who makes his first overland journey to California, for he can form no idea of the many trials he may be subjected to. The fatigues of the journey--the hardships of traversing an almost barren wilderness of nearly two thousand miles, I care but little for; but it is the narrow-minded ribaldry--the ceaseless strife which is constantly marring the tranquility of such a crowd--a mass of men in which each individual acts independent of all the rest, caring for none but himself, which renders it almost insufferable.

Alonzo Delano, Lawson's Settlement, California, September 18, 1849.

Wells Fargo stagecoaches being loaded at Cisco Station in 1867. The passenger station and Wells Fargo office were next to the main line. Until completion of the Summit Tunnels in 1868, stagecoaches and freight wagons loaded here for Virginia City and the Comstock mines.

prior to the gold discovery; subsequent to that event, thousands passed up the valley or followed up the winding course of the beautiful Truckee, stopping for a few days of rest, and to permit their jaded and half-famished stock to recruit upon the rich grass that grew so luxuriantly along the water-courses.

Upon this they feasted and fattened until spring, when they were able to easily pull their loads across the intervening mountains to the goal their now impatient masters were so eager to reach.

It will be observed that the spelling of the name of the valley was different from that in vogue at present, and is, no doubt, the more proper; the well-known faculty of the heedless miners of corrupting foreign or un-English words into a similarity to English, having, as soon as the great influx of strangers set in, changed Wassau to Washoe.

Passing by these transient sojourners, it is found that no actual settlement was made until 1852, when a man named Clark built a little cabin in a lovely spot near the present site of Franktown.

The first permanent settlement in the Truckee Meadows was made by a Mormon named Jamison, who came up from

Carson Valley in 1852, and established Jamison's Station, on the Truckee River, where he traded with the emigrants, buying their lean and exhausted stock, or trading good cattle for them. The poor cattle which he bought were quickly fattened and put in good condition by the rich grass that skirted the banks of the beautiful stream. Arriving at this station and others established on the various routes of travel, footsore and weary from their long tramp across the alkali desert, their provisions nearly exhausted, their cattle jaded and useless from overwork and lack of nourishment, the emigrants were able to lay in provisions for the balance of their journey, and to procure good cattle to replace those unfitted for further use, or whose bones lay whitening on the scorched sands of the desert.

In June, 1854, the company of Mormons, headed by Elder Orson Hyde, arrived in Carson Valley, and in the summer of 1855, Alexander Cowan and wife, now Mrs. Sandy Bowers, came to Washoe Valley, and purchased the Bowers Ranch of Dodge & Campbell. Elder Hyde was pleased with the location, and commenced the erection of a saw-mill at Franktown, as the little center of the settlement was named. In 1856 another

Extensive logging operations are conducted along the Truckee, and it is one of the sights of the trip to witness the shooting of the logs along timber-ways for one thousand two hundred feet down the side of the ridge. They make the descent in thunder and smoke, and each log, as it strikes the water, will send up a beautiful column of spray a hundred and fifty feet, resembling the effect of a submarine explosion.

Benjamin Parke Avery, 1850s

The Riverside Hotel (photo 1903) sits on the exact location where Reno began in 1859. C.W. Fuller operated a log building here that provided food and shelter to gold-seekers who were passing through the area in the reverse gold rush called the "Rush to Washoe." Myron Lake owned the property from 1861 into the 1880s, running consecutive hotel businesses under the name Lake's House. After Lake's death, his daughter and son-in-law operated the hotel and renamed it the Riverside. A subsequent owner, Harry Gosse, converted the small frame building into a lavish brick hotel, retaining the name Riverside. This version was destroyed in a fire.

Steve Crandell Collection

The few who had preceeded us on the "fools' cut-off" had paid but little regard to a definite route, they, like ourselves, only aiming now to keep a westerly course and look for grass and water, as their only hope.

A dim pack trail at this point was the only evidence that we were not the sole human beings on that desolate, treeless and almost waterless region. We filed in to the gorge with Captain Hardy in the lead. As we advanced it became more narrow and difficult, if not dangerous. The only available pack trail led along the precipitous side of the gulch, perhaps 100 feet above the bottom. In passing the point of a projecting rock, the trail being very narrow, one of the horses, losing his balance by a misstep or the pack striking the point of rock, he rolled and struggled to the bottom of the gorge. With the exception of smashing a long-used coffee pot and frying pan--very necessary articles for our housekeeping--and a few bruises on the frightened animal, no serious damage was done.

David Augustus Shaw, 1850.

Steam traction engine pulling wagon loads of lumber along the Dog Valley grade near Verdi, California, 1888

party of Mormons arrived from eastern Utah, some twenty or thirty families in all, and settled chiefly in Washoe Valley. This was quite an addition to the population of Franktown, which then became quite an important portion of Carson County, Utah, of which it was then a part.

Abandoned by the Mormons.

The recall of the Mormons by Brigham Young, in 1857, nearly depopulated the thriving community of Franktown, leaving but two ladies in the settlement, Mrs. John Hawkins, whose husband did not return with the others; and Mrs. Alexander Cowan, who refused to accompany her husband back to the home of Mormondom.

A few months after the Mormon adherents left their prosperous settlements, at the dictation of Brigham Young, fully as large a company of apostate Mormons arrived here from Salt Lake, having abandoned the City of the Saints, disgusted with its wickedness and crime.

Miners Take Possession.

Such was the condition of the Washoe Valley and vicinity when the announcement of the great discovery on Mount Davidson brought the army of miners and adventurers from California.

Thousands who came by this route passed through Washoe Valley. The little town of Franktown, with its one saw-mill, began to be of importance. Saw-mills were built in the mountains, and lumber and wood prepared in great quantities and conveyed across the valley and intervening mountains to the scene of activity. Produce of every kind from the farms, especially hay and barley, were in great demand, and more land was brought under cultivation, the yield of the farmer's toil bringing high prices.

The population of this county began to increase, and the census of 1860 showed that there were fifty-eight families and 543 people within the limits of Washoe County.

The towns of Ophir, Washoe City and Galena all blazed up in 1861, and entered upon a career of prosperity that lasted several years. Ore was hauled across the barren mountains and the marshy ground at the head of Washoe Lake, and crushed at

One plan of acquiring sudden wealth was to "salt" a wild cat claim and sell out while the excitement was up. The process was simple. The schemer located a worthless ledge, sunk a shaft on it, bought a wagon load of rich "Comstock" ore, dumped a portion of it into the shaft and piled the rest by its side, above ground. Then he showed the property to a simpleton and sold it to him at a high figure. Of course the wagon load of rich ore was all that the victim ever got out of his purchase.

Mark Twain, 1872

The Union Hotel and Saloon, believed to be in Reno

Steve Crandell Collection

Teamsters near the Yuba River from Summit Valley, Castle Peak in the background, late 1860s

Upon my arrival at the Sink, I found Mr. Moody in a much improved condition. We were unable to get any trace of the balance of our company, and determined, late in the afternoon of the second day after my arrival, to start for Carson river, 45 miles distant I packed my pony and we started at 5 o'clock p. m., and made the distance, by constant walking, in 13 hours, arriving at "Ragtown," two or three miles before reaching Carson river, at 8 o'clock in the morning. The last 15 miles of the road was a loose, yielding sand.

This had been a most disastrous piece of road to those who had preceded us. The sand was of sufficient depth to cover the wagon fellies as the jaded and worn-out animals labored under the stimulant of the brad and lash to draw their burdens. "It was the last straw that broke the camel's back." The last 10

continued in far right column

the several mills, and the teams returned with wood, lumber and produce, thus having a load both ways, and rendering the cost of getting the ore to the mill less than it would otherwise have been. The Ophir Mill of seventy-two stamps cost $500,000, and the Dall Mill, at Franktown, with sixty stamps, cost half as much. These, with the other mills in the valley, employed hundreds of workmen, and with the farms and lumber interests supported a busy population.

At the same time along the Truckee River were settlements. Within a mile of the present town of Verdi was built a bridge; at Hunter's another was constructed; at Lake's Crossing, now Reno, another; and at Stone & Gate's Crossing, afterwards Glendale, still another. At all these points did the great travel of the Henness Pass and Donner Lake routes cross the river. Stages rolled swiftly along with their crowds of passengers, while long lines of pack-trains and mule and ox-teams, drawing the capacious prairie schooner, toiled slowly along behind.

RAILROADS:

THE CENTRAL PACIFIC RAILROAD [CPRR] crosses the entire State, having a length of 433 miles within its limits, being more than half of the direct line from San Francisco to Ogden, constituting an artery of commerce upon which the life of business depends. How this power was acquired, and how it is used, a true history of its rise and operations will tell. Nevada existed and prospered before a mile of the railroad was constructed. All her vast territory was explored, prosperous and busy cities were built, elegant and powerful quartz-mills were erected, farms cultivated, the herdsman's cattle grazed upon her thousand hills, stages rattled, and the great freight wagons rolled along her interior roads, bullion flowed in a grand stream to the marts of the world, and all without help from the railroad. But this great triumph of modern art was most ardently desired. The imagination pictured untold benefits to arise from its construction. The transcontinental railroad was the great desire of the nation, and the most practicable route lay across the breast of Nevada.

Mr. Theodore D. Judah, who had been the engineer of the Sacramento Valley Railroad, from Sacramento to Folsom, California, and also of the California Central, from Folsom to Lincoln, in the same State, had explored the Sierra Nevada for routes and passes for wagon roads and the railroad, and decided upon what was known as the Donner Lake route as the most feasible.

In Sacramento were a couple of hardware merchants with whom Mr. Judah had had business relations while acting as engineer of the Sacramento Valley Railroad, Messrs. Colis P. Huntington and Mark Hopkins, and these gentlemen solicited other friends and men of influence to join, and the Central Pacific Railroad Company was formed. Leland Stanford was

Lake House and bridge at Lake's Crossing, now Reno

Courtesy of the Nevada Historical Society

continued from far left column

miles we could walk almost the entire distance upon the bodies of dead and dying animals, horses, mules and oxen, by the score, still attached to the wagons, lying in and along the roadside, in harness and yoke. Drivers, with women and children, had abandoned all to seek water and save their own lives. The stock with sufficient strength left to travel in some instances were detached from wagons and urged along, loose, before them. The ground was strewn with guns, ox chains and every kind of thing that had been abandoned. And to this day that sandy plain is covered with the bleached bones of the faithful beasts that perished on that fatal desert. By exercising due care and caution, I passed over the ground in safety with my train in 1853, with all the evidences of the terrible losses in '49 and '50 still visible.

David Augustus Shaw, 1850

T.K. Hymers Truckee Livery & Feed Stable, late 1800s.
Thomas Hymers operated the Truckee Livery Stable from
1869 to 1907, the largest stable in downtown Reno.

In the new mining regions comparatively few men are murdered for money. The greater proportion of homicides result from reckless bravado. Persons meet in saloons, bagnios and gambling places with deadly weapons upon their persons; they drink, gamble, dispute when half, intoxicated, banter each other, and at last draw out their weapons and for fancied causes alone slay each other. If one survives, when the moment of sobriety arrives, in nine cases in ten remorse comes, to escape which deeper draughts are indulged, more reckless conduct displayed until at last another quarrel with fatal results ensues.

then Governor of California, and he was made President of the company.

The railroad company wanted a subsidy of $3,000,000 from the State, and the subsidy was granted. The Supreme Court subsequently decided that the building of the road was a war measure, and the debt in its aid constitutional.

Congress extended their right to build eastward until the rails should join those of the Union Pacific coming from the East. In the meantime the great Engineer, the pioneer and organizer of the enterprise, T. D. Judah, had died in October, 1863.

Thus is presented the initiation of that stupendous work, the building of the Central Pacific Railroad from Sacramento, California, to Ogden, in Utah, most of the way through Nevada, placing the directors in the front rank of financiers, and filling their coffers with the result of labors not their own.

The estimates of their Chief Engineer as to the amount of money it would take to construct a railroad to the State line in Carson Valley by the Placerville route, a distance of 92 miles, was $7,015,568, or $76,256 per mile. To continue the same to Carson City at $59,000 per mile, would make a total cost necessary of $8,726,568 to connect the capital of the State with navigable waters.

The estimate by Mr. Judah for the Dutch Flat or Central Pacific route was, that it would cost from $12,000,000 to $13,000,000 to reach the summit of the mountains; and according to Governor Stanford's figures, $13,000,000 to make connection between the State line, eleven miles west of what is now Reno, and navigable water at Sacramento. Continue this line the remaining eleven miles to Reno, and estimate the cost of construction at $59,000 per mile-figures set by the rival company as its costs over a similar country—and the total constructing expense for the Central Pacific is found to be $13,649,000.

THE CENTRAL PACIFIC RAILROAD COMPLETED. On the thirteenth of December, 1867, the first locomotive ran into Nevada, reaching Crystal Peak from the California side. On the fourth of May, 1868, the track and telegraph were completed to Reno, and on the nineteenth of June the last rail was laid between Sacramento and that place, making railroad connection continuous between those two points. On the thirteenth of May, 1869, the golden spike was driven, and the two oceans were united by an iron band.

Central Pacific Railroad bridge over the Truckee River, late 1860s.

Library of Congress

FROM UTAH VALLEYS.
The *Placerville American* says: From Col. L.A. Norton, who has just returned from Carson Valley, we obtain much interesting intelligence from the Valleys of Western Utah. The great immigration of Mormons, of which we have made mention, heretofore as being on the way from Salt Lake, consisting of one hundred and ten families, and nearly as many wagons, with large numbers of cattle, even thousands, were within three days' drive of their places of destination -- the beautiful valleys of Wash-ho and Truckee.

Bro. Orson Hyde is erecting a new saw and grist mill in Wash-ho Valley, to be propelled by an overshot wheel on one of the mountain streams that in such numbers and great beauty are found ever full and leaping to the valleys. Both mills will be in operation in a very few weeks.

Western Standard (SF), July 6, 1856

Snow plow on the Central Pacific Railroad, late 1860s

[CPRR] Through Passenger
Rates Between Eureka and
the Following Points.

San Francisco	$45 75
San Jose	45 75
Stockton	43 75
Sacramento	41 75
Marysville	42 85
Colfax	38 10
Reno	29 00
Virginia City	32 00
Winnemucca	16 25
Battle Mountain	11 75
Elko	10 50
Ogden	31 00

DISCRIMINATION AGAINST NEVADA. From the first a system of freight and passenger tariffs was introduced that, although low enough to prevent competition by team's or stages, yet worked a serious damage to the State. It was for the interests of the company to increase its freight traffic to the utmost extent. Manufactories within the State were institutions hostile to such an increase, therefore not to be tolerated. The prices charged were governed by a rule that permitted the existence of traffic, and took for such permit the principal profits.

The citizens of the State knowing that they had rights that "white men were bound to respect," finally commenced a public agitation of the subject.

Nevada is helpless in the grasp of the Central Pacific. The Legislature has legal power to pass laws saying what shall be done by the road within her borders, but the Central Pacific has power to take fearful vengeance for any such exercise of this right as the bill contemplates. It is not good policy to exasperate this monopoly needlessly. A few years ago Washoe County compelled the Central Pacific to pay $45,000 in taxes, which the corporation did not want to pay. What was the consequence?

Within ten days wood that had been hauled from Verdi to Reno for fifty cents per cord cost one dollar to haul.

Mr. Daggett [commented] as follows:—

"Is comment necessary upon these terrible rates? Do they not speak trumpet-tongued of impositions unparalleled in the annals of railroad ruffianism? These charges have been neither known nor credited beyond the State of Nevada. When mentioned by the press they have been denied, and with threats of still greater oppressions the railroad dictators have silenced the complaints of their victims.

In favorable weather Sacramento freights were then delivered in Virginia City at one dollar and fifty cents per hundred pounds. The railroad rates are now one dollar and forty-six cents and one-half—but three and a half cents per hundred less than old teaming rates.

Yes, pack-mule competition in Nevada, of which the directors of the Central Pacific inferentially complain as a sort of wicked and unnecessary menace to their financial well-doing, is indeed all that stands between them and the establishment of rates in keeping with their rapacity, and their charges are scheduled just a shade below figures that would line the roads again with pack-trains and wagons.

Their object seems to be to crush, not to develop, the industries of Nevada."

Central Pacific Railroad freight engine, late 1860s

About noon we came in sight of the lake or sink of the Humboldt, a handsome sheet of water of some five or six square miles in extent. The water is unhealthy and avoided if possible by the emigrants though the cattle are obliged to drink it. At the southern extremity of the lake, a creek runs out of it which here is hardly fordable but in a few miles it is entirely lost in the sands of the desert. We camped near this creek some three miles from the lake. We found some old wells of a few feet in depth of water slightly brackish and filled out 2 Gal cask with the water for our cattle.

William Henry Hart, September 6, 1852

On the night in question Under Sheriff Kinkead and Deputy Sheriffs Hutton, Jones and Avery, posted themselves about town to watch the actions of five men who had attracted their attention. Sometime after midnight a shot and cries for help were heard issuing from the alley back of Commercial Row and in the rear of the post-office. Avery rushed to the scene and found the five men beating a man who proved to be W. T. C. Elliott. At this juncture Elliott fired two more shots, which, with the appearance of the officer, caused the villains to run, two going out upon Virginia Street and two upon Center. Avery pursued the first two and overtook them at the bridge, when one of them turned upon him with his gun, but when Avery covered him with his revolver exclaimed, "Don't shoot! I'm wounded now." It was found that he had a bullet wound in the right breast and another in the right leg, just above the ankle. He was taken in charge, the other man escaping.

Of the other three, one was captured by Officer Hutton, as he was escaping from the alley, and the balance made good their escape. Officers immediately went in search of them, and at five o'clock in the morning Kinkead discovered their tracks near the railroad bridge. He at once rode on in pursuit, and when

continued in far right column

J.C. Hagerman, 1874

Courtesy of the Nevada Historical Society

THE FUTURE LAND QUESTION.

From the foregoing, and from the history, so familiar to all, of the strategy, cunning and selfishness of the Central and Southern Pacific Railroad Companies in California, deductions may be drawn that portend serious troubles to the most worthy citizens of this State. The great grant of land includes much that is valuable, and much that is worthless. Alternate sections remain as Government land or have passed to individual ownership. The well-being of the State requires that all shall be utilized. Settlers are encouraged by the railroad company to occupy and improve the land, but are refused any title, or agreement of terms upon which they may rely in the future. The prospect opens before them of a repetition of the Mussel Slough War of California, with its murders, ejectments and imprisonments, its ruinous litigations, exorbitant rates for improvements made and property created by the purchaser, and at last to see one's rightful possessions owned and occupied by another. Such appears the plan and hope of the railroad corporation, ever so subtle, so far-reaching, so grasping, so powerful, and so merciless.

RENO IN ITS EARLY DAYS.

Lying at an altitude of 4,507 feet above the sea, on both banks of the Truckee River, in the rich valley so long and well known, on the old route of overland travel as the Big Meadows of the Truckee, Reno is the center of the most important agricultural district in the State, the terminus of the Virginia and Truckee Railroad, and the principal station in Nevada on the line of the Central Pacific, at which point goods destined for Carson City and Virginia City are transhipped.

In 1859 a settlement was made on the south side of the river, where the Lake House now stands, by a man named Fuller. At this point the river could be forded, and a route of travel was laid out from California, crossing the river at this point to Virginia City and the south. The house was kept as a way side inn for the accommodation and refreshment of travelers and the long pack-trains and freight teams that toiled across the mountains to the newly-discovered land of silver. This was but one of several points where the river might be crossed, and in 1860 Mr. Fuller, then proprietor of the road, upon which a franchise to collect toll had been granted, constructed a wooden bridge for the better accommodation of travel. The winter of 1862 was one of exceedingly high water, and the bridge, in common with others along the stream, was carried away by the torrent.

In 1863, M. C. Lake came into possession of the property, and rebuilt the bridge, the place becoming known as Lake's Crossing. Again in 1867 the bridge was damaged by high water and rebuilt by Mr. Lake. In 1865 an English company built the Auburn Mill, about one mile from the site of the town.

continued from far left column

he arrived at Huffakers, ascertained that they had taken breakfast there. Although he was warned that they were well armed and was advised not to attempt their capture, he continued the pursuit alone, overtaking them at Crane's. Riding up to within 100 yards of them, he dismounted and ordered them to surrender. They drew their weapons and took each a side of the road. Kinkead's shot-gun was loaded with buckshot, and covering one of them with this the officer warned him to throw down his pistol before he counted three or he would shoot. The only response to this was a laugh, and when the fatal three had been counted, the officer fired, lodging two balls in the man's right breast. They then threw down their pistols and surrendered and were safely conveyed to Reno by their plucky captor.

In two weeks they were tried and sentenced to twelve years in the penitentiary.

Milatovich Groceries & Liquor, Virginia Street, Reno, probably 1870s

Courtesy of the Bancroft Library, University of California, Berkeley

The Isaac Barnett or Thos Barnett's Dry Goods Store? The sign on top says Thos, and the sign over the door says Isaac. Reno, probably in the 1860s.

Previous to the completion of the railroads in Nevada, stage coaches were the medium of travel for passengers, and they at the same time carried all of the species into the mining towns as well as all of the bullion away from them. Large sums were constantly going over the main routes of travel, and consequently the stage coaches became an attractive feature for the more enterprising class of "bad men" with whom the State swarmed in the flush times of silver mining. Upon some of the roads these robberies became so frequent that guards were sent with the coaches, and some of the robbers soon became so well known.

This location was selected because of the good facilities for fuel and water. Quartz was brought to the mill from a considerable distance in several directions, it being the only mill nearer than Washoe City.

When the Central Pacific Railroad began ascending the mountains with giant strides, the officials looked ahead of the iron horse to select suitable spots where should be established the necessary stations. Somewhere on the Truckee River it was evident must be a point where the goods for Virginia City and vicinity would be unloaded and forwarded to their destination. It was well understood that such a town as that was destined to be of considerable importance, and care was used to select the most eligible situation. The land on which the original town was laid out belonged then to M. C. Lake, and the only building upon it was one he had erected at the north end of the bridge with the intention of building a grist-mill.

He deeded forty acres to Charles Crocker in consideration of his causing a station to be established there, laying it off in town lots, and conveying a certain number of the lots back again. This was accordingly done. The town was christened Reno, in honor of General Jesse Reno, who fell at the battle of South Mountain.

As in all such cases since the palmy days of '49, saloons were the first places of business to be opened, and by far the best patronized. For a few weeks men had nothing to do but to see to it that these "necessary evils" did not fail for lack of an occasional two-bit piece, and it is hardly necessary to remark that they attended to this duty with a zeal worthy of a better cause.

The first train from Sacramento arrived June 18, 1868, and it was a great day for Reno, bringing with it the tangible assurance that their confidence had not been misplaced. From that day Reno counts its career as a business town. It was nearly a year after this event, on the tenth of May, 1869, that the last spike in the overland railroad was driven at Promontory Point, and a few days later the citizens of Reno assembled at the depot to greet the first through train from the East.

No sooner did trains begin to arrive from Sacramento with their loads of freight and passengers than Reno began to bustle and hum with life and activity. Stages left daily for Carson and Virginia, crowded with passengers, and long trains of freight wagons were loaded with goods at the depot, from the scores of cars that arrived weekly, and defiled through the streets and out upon the roads that led to their destination.

Those were the palmy days of Reno; work for all who sought it; plenty of money; good prices paid for labor and goods. The number of men, animals and wagons required in transacting this immense freighting business, assured employment and prosperity for the merchant, farmer and mechanic, and, it may be remarked, to the saloon keeper also. Where there is a large number of men, well employed and receiving

July 4. [1864] Charles H. Plum was stabbed and killed at a ball in Ophir, Washoe County, by a brother of a girl he kissed in a sportive manner, when dancing with her.

Clark & Wollam, Blacksmiths, Horse-shoers & Wagon Repairers. Reno, August 1893

Reno Blacksmith, Oar Burke

On the thirty-first of January, 1871, the stage running between Reno and Honey Lake was stopped by two highwaymen not many miles from the former place. The driver, a Mr. Thomas, who was also owner of the stage, was ordered from the box and relieved of ninety dollars. In the stage as a passenger, was Major Eggleston, "United States Army Paymaster, who had in a purse in his pocket two hundred dollars; and a belt upon his person containing seven thousand dollars in currency, all of which the robbers took. They also took a small sum from another passenger. While the robbers were engaged with these last two, and off their guard, Mr. Thomas pulled out a derringer which he had in his pocket and which the robbers had overlooked, and fired at the man who had Major Eggleston's belt, causing him to drop it. The shot wounded the man, but the two opened fire upon the passengers, meanwhile retreating. During the melee the horses ran off

continued in far right column

good wages, especially when the majority of them are unmarried and free from the restraining care of the home circle, there the saloon finds its most inviting field. There, also, will be found a class of human cormorants who live upon the labor and toil of others by robbing them at the gaming table, or by the many devices of which money is extorted from the unwary, or, failing in that, by open violence and crime. With such a class, in common with her sister towns, was Reno infested. Saloons and gambling houses opened their inviting door, and shameless women walked the streets and enticed men into dance houses where music and revelry sounded far into the night. Such were the infant days of Reno.

In 1871 L. H. Dyer built a theater, and thus added one more metropolitan feature to the town.

In September, 1872, connection with Virginia City by means of the Virginia and Truckee Railroad was completed, and Reno realized that what had been looked forward to as a great advantage, was, for the time being, a severe blow to its prosperity. Indeed, the citizens had begun to see this sometime before, for the year before the road had been built from Reno to Steamboat Springs, and the latter place, for the time, became the terminus of the road and the point where goods were transferred to wagons to be carried to their destination. The immense freighting and stage business to the south that had kept Reno bustling with activity was transferred to Steamboat Springs, and upon the completion of the road died out entirely.

INFESTED BY BAD CHARACTERS.

Mention has heretofore been made of the number of bad characters that infested Reno. Several times the city had been nearly cleared of them by means of notices sent to the more notorious ones to leave within a stipulated time, signed "601," and known to eminate from an association of citizens.

For some time prior to the thirteenth of July, 1874, Reno had been made the rendezvous of three-card-monte-men, gamblers, garroters, and burglars. Men had been fleeced of their money, houses and stores entered and people robbed on the streets.

THE ASSOCIATION OF "601." In July, 1874, there was formed an Association of citizens, who were known as the "601." The object of this Association was to find out and watch any objectionable characters that might infest the town, and to give them "tickets of leave" whenever it was deemed necessary to rid the town of their presence. These notices to quit the place were often more effective than suits at law or open violence would have been.

The only time that it became necessary for the "601" to demonstrate the fact that they were not a mythical organization was in September, 1878.

There lived in town at that time a saloon keeper, named W. J. Jones, whose unsavory reputation had followed him hither

continued from far left column

with the wagon and went to Reno alone, whereupon a number of persons came out to learn what had happened. Chase was given the robbers and the wounded one was caught.

the following year, August 16th; the same stage was stopped a mile and a half from Reno by three armed men who sprang into the road by the side of which they had been concealed, and ordered the driver, Mr. Thomas, to stop. Instead of doing so he whipped up his horses and the robbers opened fire. A wounded horse soon caused the stage to halt when the robbers came up and completed their job, but not until the driver and his three passengers had ineffectually exhausted all their shots in an endeavor to keep them away. For this Jackson Morrison and Clement B. Lee were sent to prison.

Reno view east of the Central Pacific Railroad

Courtesy of Bancroft Library, University of California, Berkeley

Trail Drivers in Nevada, probably in the 1870s

I can remember no night of horror equal to my first night's travel on the Overland Route. An American friend, who had himself crossed the plains, had recommended me to bring an air-pillow. This became my mainstay: I sat on it by day, or interposed it between the hard side of the coach and my ragged skin and jaded bones, and by night I put my head through the hole in the middle and wore it as a collar, like a degraded Chinaman. This saved the sides of my head during my endeavours to sleep, but occasionally a heavier jolt than usual would strike the cranium violently against the roof, driving it down between my shoulders.

Edmund Hope Verney, 1866.

from California. A young lady in San Francisco inserted an advertisement in one of the papers, seeking for a situation as a lady's companion. Jones answered it, and stated that he was an invalid lady, and finally made arrangements with her to come to Reno. "When she arrived here, late at night, Jones met her at the depot and conducted her to his saloon, the character of the place not being observed by her until she had entered. Here he made insulting proposals to her, which she resented, and compelled him to conduct her to a hotel. A companion of Jones, one H. J. Carson, then went to the hotel, and by representing to her that she was not safe from Jones there, induced her to accompany him, to what he called, a place of safety. He conducted her towards the railroad bridge, and then made the same overtures that she had received from Jones. The now thoroughly frightened girl, alone and friendless in a strange place, and at the mercy of such villains as these, knew not what to do or which way to turn, but finally reached the hotel again, and related her story.

The indignation of the citizens was intense when the news was circulated the next day. Carson was arrested for vagrancy and lodged in jail, much to his satisfaction, for he feared the vengeance of the people. The young lady was taken in charge by the Masons, and tenderly cared for.

The next evening, September 19, 1878, Jones was visited in his room by a body of men, who bound him and carried him to the south end of the railroad bridge. That evening a much respected citizen, William Duck, had died, and as the captors proceeded with their victim the church bell was tolling. Imagining that the bell was sounding his own death-knell the guilty wretch begged and pleaded for mercy.

Arriving at the end of the bridge they found more men who had in charge a large kettle of tar and a liberal supply of feathers. He was deprived of his clothing, covered with the hot tar, a kettle of the hot liquid emptied over his head, his face, hair and eyes literally filled with it, and then liberally covered with snow-white feathers. His clothing was then put on him, and trembling with pain and fright he was given a ticket to Truckee, and placed on the Overland Train.

On the train and in Truckee he was the subject of a great deal of pity by people who were not conversant with the facts, and the act was denominated a cruel outrage by the newspapers. When, however, the circumstances were brought to light he received but little sympathy. He was several days in Truckee before he became thoroughly cleansed, and the blisters made by the hot tar were a constant reminder of the "601" of Reno for many days. Carson was sentenced to fifty days in jail for vagrancy, and when discharged took his departure from town.

End of track, for that day at least, near Humboldt Lake. Workers demonstrating their jobs during a visit by dignitaries, 1860s.

November 16. [1879] S. M. Oakes was shot and killed by Mrs. Dr. Snow, at Reno, Washoe County. Oakes went to the house after the doctor, and, being deaf, did not hear Mrs. Snow's question of "Who is there?" She supposing him to be a burglar, fired a shot through the door, which killed him. It was a deplorable accident.

Library of Congress

... even during the flush times that "grass was short," so it chanced that the two concluded to reduce expenses by setting up what they called a bachelor's hall. A comfortable little cabin was rented on the outskirts of Ophir for the winter of 1864, and their worldly goods and chattels moved in. . . . all went well until they discovered that the supply of wood was about exhausted. Wood was plentiful not far away, but they had no wagon with which to bring it. A short consultation resulted in their going to Jim Sturtevant and asking him to haul them a few loads. Mr. Sturtevant demurred and said: "Boys, I am as lazy as you are; haul your own wood." They explained they had no team. Sturtevant then told them to take his two yoke of cattle and haul all they wanted. This matter being arranged, Judge Healey and Jim Gatewood started up the canyon, Jim doing the driving while the Judge held down the wagon. All went well going up hill. The wagon was soon loaded and the team headed down the grade, but here trouble commenced. The wagon crowded the wheel cattle so that the team jack-knifed and an upset was imminent. But the oxen were finally halted and it was then arranged that the Judge should take a position on the off-side of the cattle and assist in keeping the team in the middle of the road. When all was ready Jim admonished the "damn bulls" to act decently and they started down the road. But the cattle were nervous. The outfit got going faster and faster until it was evident that dire destruction was sure to come. Jim got excited and finally yelled out to Healey, "Stop them, Judge, stop them; why in damnation don't you stop them?" This profanity was too much for the Judge, so he stopped short and yelled back to Jim: "Stop them yourself, I am no damn bull driver! I am a Kentucky gentleman, sir !"

Chinese camp on the Central Pacific Railroad at Brown's Station, 1860s

Notices were also sent to a number of undesirable citizens to take up their abode in some remote locality, and some of them departed without even waiting for this little formality. One of these, a young man named Alf. Howard, or better known as Jesse Cook, had the temerity to return on the twenty-eighth. He had made himself obnoxious by circulating obscene literature, and enticing drunken men into houses of ill-fame, and the "601" determined to show that they meant what they said when they issued an order to leave town.

About seven o'clock the next evening he was enticed into the alley back of Morris Ash's saloon, where he was seized and bound by a body of men. His cries for help brought a number of people to the rescue, who departed as hastily as they came when permitted to gaze into the muzzles of the numerous revolvers carried by the men. Cook was taken to a secluded spot on the river bank, and given a very light coat of tar and feathers on his face only, being treated leniently on account of his youth. He went to Truckee and joined his father, who had previously been driven from Reno. They returned the next morning, and took the train for Virginia City. Since these events it has been unnecessary for the "601" to make any demonstrations whatever.

PRINCIPAL TOWNS AND CITIES.

CRYSTAL PEAK lies in a grassy nook, between the jutting hills at the eastern foot of the Sierra. It is the natural outlet for an extensive tract of timber land, and for that reason, and because of its beautiful and healthful location, a splendid site for a thriving town. The advantages were noted and improved by the Crystal Peak Company, who laid out a town here in 1864. The company owned lumber and mining interests some ten or fifteen miles west of the town, in a mountain containing crystallized gold quartz, from which the name was derived.

Several saw-mills in the vicinity have been running constantly; the Truckee is capable of furnishing an abundance of water-power, and for years the lumber and wood supplied from this district were sufficient to support quite a flourishing town.

In the year 1868, Crystal Peak enjoyed a prosperity such as she has not known since. The Central Pacific Railroad was then just entering this State, and the saw-mills of this section were supplying the immense amount of necessary materials for its construction. All was bustle and business in the little town, and hopes were entertained that the road would pass through it, but they were not realized, for the line passed two miles to the left.

FRANKTOWN is the pioneer town of Washoe County, being settled before the great Comstock Ledge was discovered and

"Towards morning there was a commotion among the passengers. A sudden shock roused all from their slumbers. Many were greatly frightened, but no one was seriously hurt. A severe shaking was the only result of what proved to be a collision with a herd of cattle. The engine and tender had been thrown off the rails. Two oxen were crushed to death... a detention of eight hours between Wadsworth and Clarks' Station and the loss of breakfast were the only sufferings to be borne.

An hour did not pass away before two locomotives were on the spot. What was still more important, the passage of trains over the line was stopped... Some passengers were indisposed to forego their breakfasts without an effort to provide a substitute. There was plenty of beef alongside the line, and the sage-brush could be used for fuel. What more natural then, they argued, than to light a fire and cook a steak? ...the fire threatened to produce serious consequences. The flames rushed along in the direction of the telegraph posts and the cars. A German gentleman of greater pluck than prudence had ignited the sage-brush, and he became ludicrously alarmed at the results of his act. He rushed about in frantic consternation, making energetic attempts to stamp out the flames. His vigour in undoing the mischief he had caused, led to the scorching and permanent injury of his boots and trousers.

William Fraser Rae, 1869

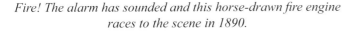

Fire! The alarm has sounded and this horse-drawn fire engine races to the scene in 1890.

Steve Crandell Collection

A band of robbers, led by A. J. Davis, and composed of J. E. Chapman, R. A. Jones, E. B. Parsons, John Squires, James Gilchrist, Tilton P. Cockerill and J. O. Roberts, planned the robbery of Wells, Fargo & Co's, treasure-boxes. Chapman went to San Francisco to watch for a large shipment of treasure, which was expected soon to be made. On the fourth of November, 1870, he sent the following dispatch by telegraph:—

"To R. A. Jones, Capital House, Reno Send me sixty dollars, and charge to my account" J. ENRIQUE.

This dispatch was conveyed to a retreat in the Peavine Mountains, in which were assembled all of the gang except Chapman and Roberts, and where the plans for the robbery were matured.

That afternoon they all proceeded to the stone culvert, near Hunter's, going by different routes. Here Jones was left with the guns and tools, with the understanding that soon after the freight train passed up to Verdi the others would be down with the engine and express car, and that if they did not stop at the culvert, to place obstructions on the track to prevent pursuit, and to follow on his horse with the guns and tools.

Davis, Parsons, Squires, Cockerill and Gilchrist then proceeded to Verdi, and when the eastward-bound train

continued in far right column

J. King's wool wagon.

before there existed such an organization as Washoe County, as has been fully detailed in the preceding history of the county.

The town of Franktown was first settled in 1852, and became a town in the year 1855. Its early history has been fully related in the history of the settlement of the county, with which it is too closely woven to be separated. It was but a small hamlet, and was the only town within the present limits of the county of Washoe, until after the influx of people caused by the silver excitement. The sawmill built by Orson Hyde was the only manufacturing industry, the settlers being nearly all farmers.

The discovery and development of the Comstock soon had an effect on Franktown. The saw-mill was run to the limit of its capacity, as were others in the vicinity. A brisk demand for wood and lumber for the mines kept Franktown busy, as it did other places in the valley. A sixty-stamp quartz mill, costing $250,000, was erected here, by J. H. Dall & Co., in 1861, and caused a great increase in the business and population of the town. It was burned in 1865, and immediately re-built, but was a second time burned, a few years later.

In 1872 the Virginia and Truckee Railroad was completed, and this place became quite a depot for the shipment of wood, lumber, and produce, from the surrounding farms and the timber lands in the adjacent mountains. There are a hotel, two stores, market, blacksmith shop, and a number of neat dwelling-houses. A wood flume, owned by the Virginia and Gold Hill Water Company, terminates sit this point.

GLENDALE was formerly known as Stone & Gates' Crossing, a trading-post having been established here in 1857, by Charles C. Gates and John F. Stone. It is but a few miles below Reno, find a portion of the travel to Virginia City crossed the river at this point, instead of at the several crossing places above. Stone & Gates kept the Farmers' Hotel at this point. In 1860 Stone & Gates built a bridge here, which was carried away by the high water in 1862, when the county constructed a free bridge. A store was built here in 1866, and soon quite a town sprang up, consisting of two stores, hotel, market, blacksmith shop, saloons, etc., which received the name of Glendale. It enjoyed its lease of life but a short time, however, for two years later, the new town of Reno absorbed all the business it formerly enjoyed, and the town of Glendale vanished from sight.

GALENA was laid out in the spring of 1860 by A. J. and R. S. Hatch, who then organized the mining district of Galena in the edge of the mountains on the west of Pleasant Valley. They also built a smelting furnace, the first one on this side of

Crystal Lake, altitude 5,907

Library of Congress

continued from far left column

stopped there, about 1 o'clock on the morning of the fifth, boarded it, cut off the passenger coaches, took possession of the engine, mail and express cars, and compelled the engineer to run down the track and stop at the culvert. Davis then cried out "Man, come out with those guns," when Jones made his appearance. The door of the express car was then opened, and the messenger ordered out and placed under guard with the fireman in the mail car, a guard being also maintained over the engineer. The treasure-boxes were then broken open and $41,600 secured.

Having accomplished the robbery, the men hastily divided the plunder, and departed in different directions. Davis went towards Virginia City, burying $20,000 near Hunter's place. Jones and Gilchrist went across to the Peavine road, with $7,500, which they buried in a ravine near a point of rocks, and continued on to Sierra Valley, where they were soon after arrested. The others, with the balance of the spoil, took the road to Crystal Peak, scattering in several directions, one of them going to the house of J. C. Roberts, in Antelope, another member of the gang.

Within a week after the commission of the crime, the perpetrators were all arrested, including Chapman and Roberts, some

continued far left column of next page

continued from last page far right column

of them in this State, and others in California. Roberts confessed all he knew about the affair. Jones divulged the hiding place of $7,500, Gilchrist of $12,000 and Davis of $20,000, as that nearly the whole amount was recovered.

At the trial in Washoe City the following month, Roberts and Gilchrist testified against their companions and were discharged. Davis and Jones pleaded "guilty," and were sentenced, the former to ten years and the latter to five years, in the penitentiary, while the others pleaded "not guilty" were convicted and sentenced to various terms, ranging from eighteen to twenty-three and one-half years. In what is denominated the "Big Break" from the penitentiary, September 17, 1871, in which twenty-nine prisoners escaped, Squires, Chapman, Parsons and Cockerill gained their liberty, but were all recaptured within a month. Parsons was captured September 28th, and confined in the Ormsby County jail, from which he immediately escaped, and remained at liberty several years.

Ophir City, 1870

Courtsey Library of Congress

the Sierra, and constructed a road one and one-half miles long from the town to the mines at Galena Hill.

The district received its name from the large quantities of galena in the ore. Several unsuccessful trials were made to reduce the ore. The ore was too base and the amount of silver too small to be worked to advantage, and the mines were abandoned.

At this time the business of the town underwent a radical change. The town was moved half a mile further up the creek, and it became a flourishing lumber camp. For five or six years the business was good, and the town had a population of over 300, chiefly Italians; but as soon as the lumber became exhausted the town disappeared.

HUNTER'S BRIDGE is a crossing point of the Truckee, midway between Reno and Verdi. It was on one of the routes of travel to the Washoe country from California. In 1860 a man named Stout built a bridge here. John Hunter also kept a hotel at this place. In 1862 Mr. Stout was drowned, and the bridge carried away by high water, but the Henness Pass Toll-road Company rebuilt the bridge, which became free upon the expiration of the franchise in 1872.

JONESVILLE was laid out two miles from Pyramid City, at which point is situated the Jones & Kinkead Mine, the most important in the district, and the one on which the most work has been done.

MILL STATION is two and one-half miles south of Franktown. This was an old mill-site; and is now the terminus of a wood flume from the mountains, and a station on the Virginia and Truckee Railroad, and contains several little cabins.

OPHIR is three miles below Washoe City and one mile above Franktown. Here the Ophir Mining Company erected a quartz mill and reduction works in 1861. To this mill was drawn all the ore taken from the company's mine in Virginia City.

PYRAMID DISTRICT lies a few miles west of the south end of Pyramid Lake. As early as 1860 prospectors were through this region, and ledges were discovered, but were considered of little value and were not worked. The croppings along the surface are exposed to view for a long distance, and lay unnoticed for a number of years. On the sixth of March, 1876, Dr. S. Bishop, of Reno, located the Monarch and was soon followed by many others.

A two-stamp prospect mill was erected by Bishop, and the result of its workings caused quite a rush of people to the new district. The ore so closely resembled that of the Comstock that it was proclaimed that "another Comstock" had been found, and some went so far as to assert that it was the same vein as its noted predecessor of Mount Davidson.

It was interesting to note that the stock was about equally divided between horses, mules and oxen. With the exception of getting footsore, oxen appeared to stand the journey about as well as horses or mules, while all were eaten in emergencies. Oxen, even when they died from hunger and fatigue, were preferred, except by the Indians, who had no choice.

David Augustus Shaw, 1850

Miners pose with pails and a lantern near a mine building of the Ophir Shaft near Virginia City, Nevada, late 1800s. Although the photo caption labels the pails "water" pails, their size and shape suggest they may have been lunch pails.

Denver Public Library, Western History Collection

October 20, 1849

Affectionate Companion

The prospects for obtaining gold are not as good as we had been led to anticipate, though it may appear better when the rain begins to fall We have dug but very little as yet having been prospecting and looking after our Cattle and building some log shanties for the winter &c. &c. This is one of the poorest rich countries that I ever heard of-So for as I have seen there is scarce an acre of land that is tillable. The summer drouth is alone sufficient to prevent almost every thing from growing (no water falling between the months of March and November) but there is but little that would be called good were this objection obviated

E.A. Spooner

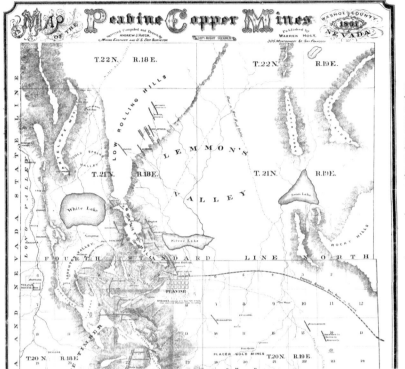

Peavine Copper Mines Map, 1867. See page 116 for larger view.

Pyramid City was at once laid out, and a boarding-house and a few buildings were erected, the population soon amounting to nearly 300. During the summer of 1876 daily crowded stages ran from Reno to Pyramid City.

POEVILLE, a small mining camp, sometimes denominated Peavine, Poe City, or Podunk, is situated in the Peavine Mountains in the Peavine Mining District, nine miles from Reno.

The Peavine ledges were discovered in 1863, and a district twenty miles long and about ten miles in width was organized. In the center of the district is a cluster of small springs, near which a house was built in 1860, and from the peavines growing about the springs the mountains received their name.

The lodes of the district are from three to twenty feet thick, and lie in a granite and metamorphic formation. Several tunnels were run in on a number of the ledges soon after the discovery of the district, and ore taken out that assayed from fifteen to forty per cent, copper, and from $60 to $500 per ton in gold and silver. Some choice ores reduced at the Auburn Mills yielded 100 ounces of silver to the ton.

The cañons in the mountains were worked for placer gold, when plenty of water was to be had in the spring.

In 1863, John Poe and others resumed work in this district, and developed several rich ledges, the Poe, Paymaster, and Golden Fleece, being the most prominent.

The ores were found to be very rebellious, some of them possessing the most complicated combinations of minerals known. Many new processes were introduced and tested here, each one with a great deal of confidence, but all to no purpose; and, although the ores assayed extremely high, enough could not be extracted from them to pay for the working.

STEAMBOAT SPRINGS. Nature, in an eccentric mood made these springs for the benefit of mankind, and in this, as in others of her wonderful creations succeeded admirably. They are situated in Steamboat Valley, an extension of the Washoe Valley, at an altitude of 4,500 feet above the sea, eleven miles south of Reno. The buildings consist of a fine hotel, with

While resting at noon I interviewed a Pike county Missourian, who "allowed" he could spare a "right smart piece of bacon." I asked him the price. "I reckon about two bits a pound; it's a doggone long ways to haul it stranger." I replied that I would be willing to pay twice that amount rather than not have it. "Wouldn't take a cent more, stranger, not a cent more. Didn't cost me more than two bits to haul it here; wouldn't take a cent more."

David Augustus Shaw, 1850

Ophir Mine, Virginia City, probably 1860s

Train crossing Truckee Meadows, late 1860s

The desire for gold has often led to many a fatal snare, an cankered many an, otherwise, noble spirit; and after all, perhaps they have not obtained it: or if they have, they have not always been able to keep it, and have found themselves forsaken by the god of this world; and what is more to be regretted, forsaken by the God of the world to come. Then, to sooth sorrow and disappointment, comes the dram; next gambling, with all its shady branches, to retrieve a lost fortune. Then the last lingering ray of virtue is basely exchanged for a free and full indulgence in the lowest vices.

Orson Hyde, 1856
From Carson City, for the Western Standard

twenty rooms, also five cottages containing a like number of rooms. Connected with the main hotel is a bath-room building, containing fifteen seperate sets of baths each, a set consisting of a steam bath from a hot sulphur spring, also tub and shower baths. No artificial agencies are employed in the heating of the water.

WADSWORTH is at an elevation of 4,077 feet above the sea level, and is one of the lowest points on the line of the Central Pacific Railroad in this State. The Big Bend of the Truckee is a place familiar to all overland emigrants who came by this route, as being the place where they first found plenty of good, pure water upon emerging from the desert; and here, where the river turns to the north to find its home in the bosom of the Pyramid and Winnemucca Lakes, is where most of the emigrants reached and crossed that stream at what was known as the Lower Crossing, now called the town of Wadsworth.

It was here that Fremont left the river and continued south in January, 1844; and it was here in the fall of the same year that the party of emigrants first saw and named the river. It was one of the great landmarks of overland travel, and the one most looked forward to for its refreshing supply of water, grass and fish. It, as well as other points on the river, was a great recruiting station for exhausted emigrant trains.

When the Central Pacific Railroad passed through here, in the summer of 1868, this point was selected for one of the most important stations on the line. It is here that the road leaves the river and strikes out across the Great Desert, through which it runs a distance of 100 miles, to the town of Humboldt. The car shops of the Truckee division, extending from Truckee to Winnemucca, were located here; and here the engines take their load of wood and water for their long trip across the arid desert. For this reason Wadsworth first came into prominence, being the base of supplies for the building of the road across the desert. The engines on this portion of the road are constructed with increased capacity for carrying water, on account of the great quantity required.

The work shops at this point employ quite a number of men, and the round-house contains twenty stalls. Besides the railroad interests there are two hotels, three grocery stores, two general merchandise stores, one variety store, and saloons, markets, shops, etc.

The excellent bridge that spans the Truckee River at Wadsworth was constructed in 1879, by the county, at an expense of $4,000.

A number of accidents, of a more or less serious nature, have occurred on the railroad in the vicinity of Wadsworth, but probably the most peculiar one and the one that but narrowly

If the Americans were not the most modest people in the world, they would before this have made more famous than any European public work the magnificent and daring piece of engineering by whose help you roll speedily and luxuriously across the Sierra Nevada from Ogden to San Francisco. But we Americans have too much to do to spend our time in boasting.

Charles Nordhoff, 1872

Wadsworth, Nevada, on the Truckee River

Library of Congress

Truckee River at Verdi, 1860s

That Knox had been murdered was positively ascertained. Indians passing the former camping-ground of Vail in the cañon discovered a saddle, that had been buried, partly exhumed by coyotes. Pulling it out, they carried it to the settlements, and related the circumstance. The people having before this suspected foul play, went to the spot where the saddle was found, guided by the Indians, for the purpose of making further examinations. Upon digging, they soon found the body of Knox, who had been killed by a blow on the head, apparently with an axe, and, doubtless, while asleep. Vail had buried the body, and then made his bed over the spot, so as to hide it. This position he had occupied for more than a month—sleeping upon the grave of his victim!

escaped being most horrible in its consequences, occurred June 13, 1872. Passenger train, No. 1, passed over a broken rail, six miles west of the town, which caused the rear two coaches to leave the track and lean up against the rocky side of a cut, through which the train was passing. In this position they were dragged rapidly along until they came to the end of the cut, opening out upon a steep embankment when the two coaches were upset and demolished. Strange as it may seem when one contemplates the nature of the accident, no one was killed, but twenty-seven passengers were injured, some of them severely. Had the cars leaned in the opposite direction, they would have been deposited in the Truckee River as soon as they cleared the cut, and a great loss of life would have necessarily ensued.

VERDI is a station on the Central Pacific Railroad about two miles from the old town of Crystal Peak, springing up as soon as the railroad came along, and may be called the descendant and successor of that town. Here the lumber interests of the district are centered; saw, lath and shingle mills here find a shipping point. Of late years a new industry has been added, and now Verdi also stores and ships large quantities of pure mountain ice, being one of the points for the preservation of that article, the whole ice business of the coast being concentrated in the Sierra, along the line of the railroad. The population is about 200.

The bridge that spans the river on the road from Verdi to Crystal Peak was built in 1873, partly by the county and partly by the citizens. In 1860 a bridge was built there, the place being known as O'Neil's Crossing, and being one of the crossing points of the Truckee River for travel to the then new mining region of the Comstock. In 1862 the bridge was carried away by high water, and was rebuilt. A most curious accident occurred here on the thirty-first of March, 1873. The bridge, upon which tolls were then collected, fell into the stream, while a load of wood, drawn by ten oxen, was upon it. But one ox was injured, and the wagon was drawn out of the water, right side up, without having lost a single stick from its load. The new bridge which was then constructed was made free to all.

I encamped for the night, having to go some distance from the road to find feed and water. During the following forenoon I purchased a small amount of beef ribs from a man who was dressing a dead animal by the roadside, but whether it had been killed or had died a natural death I did not know, as no questions were asked.

David Augustus Shaw, 1850

Central Pacific Railroad alongside the Truckee River, 1860s

Library of Congress

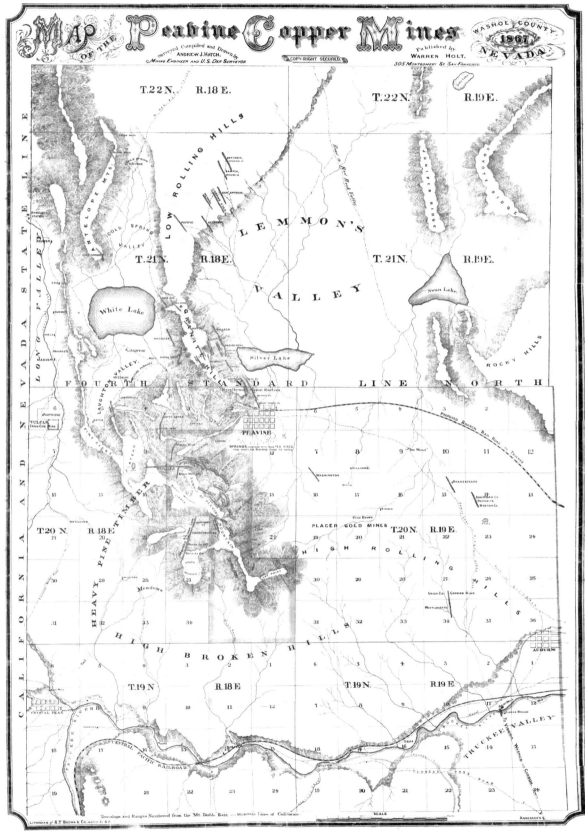

Shoshone Indians and tepees, taken sometime between 1880 and 1910. Some Western Shoshone groups, especially in northeastern Nevada, had developed cultural features similar to the Plains, including the horse and the teepee, before the Euro-Americans arrived in Nevada.

Chapter 6: Indians and their wars in Nevada.

The first intercourse between the white and red race in Nevada, of which there is any record, dates from 1832. In August of that year Milton Sublette reached the head-waters of the Humboldt River, with a company of trappers Within a few days after their arrival at that place, [Joe] Meek shot and killed a Shoshone Indian. The famous mountaineer, N. J. Wythe, who was also of the party, asked the trapper why he had done this, and was told that it was only a hint "to keep the Indians from stealing their traps."

"Had he stolen any?" queried his questioner.

"No," replied Meek; "but he looked as if he was going to."

Cady was riding along a trail not far from where Dayton now is, and overtook an Indian, and like a brave man, deliberately shot him.

Winnenap will not any more. He died, as do most medicine-men of the Paiutes. Where the lot falls when the campoodie chooses a medicine-man there it rests. It is an honor a man seldom seeks but must wear, an honor with a condition. When three patients die under his ministrations, the medicine-man must yield his life and his office. Wounds do not count; broken bones and bullet holes the Indian can understand, but measles, pneumonia, and smallpox are witchcraft.

It is possible the tale of Winnenap's patients had not been strictly kept. There had not been a medicine-man killed in the valley for twelve years, and for that the perpetrators had been severely punished by the whites. The winter of the Big Snow an epidemic of pneumonia carried off the Indians with scarcely a warning; from the lake northward to the lava flats they died in the sweat-houses, and under the hands of the medicine-men. Even the drugs of the white physician had no power.

Mary Hunter Austin, 1903

Heebe-tee-tse, Shoshone Indian, date unknown

This was a suggestive introduction of the whites to the natives of Nevada; one that gives the chief actor a distinction over which it requires, upon our part, a great effort to become enthusiastic.

The following year Captain B. L. E. Bonneville started an expedition of forty men under Joseph Walker, from the Green River Valley, to explore and trap the country west from Salt Lake to the Pacific Ocean. The company made its way slowly down the Humboldt, trapping as it went, until the curiosity of the natives had gradually overcome their fears of the whites. From day to day their numbers increased in the vicinity of, but at what they considered, a safe distance from, the camp and line of the strangers' advance. At night the more daring would occasionally steal into camp and carry off some trifling article that seemed to them a treasure of priceless value.

Their petty larceny proclivities, combined with their constantly increasing numbers, eventually aroused the suspicion of Walker, who claimed, as justification of what followed, to have feared a meditated attack.

Washington Irving, in his account of this expedition, says:

"At length, one day, they came to the banks of a stream emptying into Ogden's River (Humboldt), which they were obliged to ford. Here a great number of

Shoshones were posted on the opposite bank. Persuaded that they were there with hostile intent, they advanced upon them, leveled their rifles, and killed twenty-five of them upon the spot. The rest fled to a short distance, then halted and turned about, howling and whining like wolves and uttering the most piteous wailings. The trappers chased them in every direction; the poor wretches made no defense, but fled with terror; neither does it appear from the account of the boasted victors, that a weapon had been wielded or a weapon launched by the Indians throughout the affair. We feel perfectly convinced that the poor savages had no hostile intention, but had merely gathered together through motives of curiosity."

After the departure of Walker's party, there was no more slaughter of Indians for the ensuing seventeen years, although numerous expeditions passed through Nevada, culminating in 1849-50 in a tidal wave of whites from over the plains that passed down the western slope, a deluge upon the golden plains of California.

The passage of emigrants through the country, among whom were many that were reckless, and some who thought that the reputation of having killed an Indian would transform

Sego, Shoshone Indian, date unkiown

Courtsey of the Library of Congress

One of the party related the circumstances as follows: "I was standing near some sage bushes when I heard a rustling among them, and going in the direction of the noise, saw an Indian creeping along, who, seeing that he was discovered, let fly an arrow that just grazed my ear. He then gave a whoop and ran, but tumbled down before he could draw another arrow from his quiver. One of the boys coming to my aid and having a hatchet in his hand, rushed forward and buried it in his skull, killing him instantly. The whoop of the now dead Indian brought the whole force, and a shower of arrows fell among us. I was the first to answer with a rifle shot which brought one of the foremost savages off his horse to the ground. In the meantime my companions were using their rifles to good effect. We were able to get behind a row of willows that afforded us some protection from the arrows of our assailants. After firing the second volley of rifle shots the smoke cleared away and I could see we had made more than one savage bite the dust. I had my eye on an old man who leaped from his pony and took in his arms one of his wounded companions who had been shot through the leg. Placing him on a horse, he mounted his own led the other and rode away. I could easily have shot him, but when I saw that, I could not find it in my heart to do so, but let the old chief carry off his wounded comrade in safety. As we emerged from our shelter, all that could be seen of them were five dead ones, weltering in their blood, bows and arrows and a few feathers and tomahawks lying on the ground."

David Augustus Shaw, 1850

GENERAL CLARICE, U.S.A., Commander of the Pacific Department,

SIR: We are about to be plunged into a bloody and protracted war with the Pah-Ute Indians. Within the last nine months there have been seven of our citizens murdered by the Indians. Up to the last murder we were unable to fasten these depredations on any particular tribe, but always believed it was the Pah-Utes, yet did not wish to blame them until we were sure of the facts. On the thirteenth day of last month, Mr. Dexter E. Demming was most brutally murdered at his own house, and plundered of everything, and his horses driven off. As soon as I was informed of the fact I at once sent out fifteen men after the murderers (there being snow on the ground they could be easily traced), with orders to follow on their tracks until they would find what tribe they belonged to; and if they would prove to be Pah-Utes, not to give them battle, but to return and report, as we had, some two years ago, made a treaty with the Pah-Utes, one of the stipulations being that if any of their tribe committed any murders or depredations on any of the whites, we were first to go to the chiefs and that they would deliver up the murderers or make redress, and that we were to do the same on our part with them. On the third day out they came onto the

continued in far right column

Dat So La Lee, one of the most famous Indian basket weavers and one of the most famous Washoes, 1899

them into heroes, resulted in the slaughter of some straggling Shoshones, along the Humboldt in 1849. Several instances of the kind occurred, where they were shot in retaliation for real or fancied aggressions. In 1850 this tribe, or portions of it, commenced a series of depredations that lasted until the close of 1863.

In June, 1850, a train from Joliet, Illinois, among whom was Capt. Robert Lyon, who relates that while camped at a point near where Elko now is, they lost one of their party, who was shot through the heart with an arrow while on picket duty. An ineffectual attempt was made to stampede the horses, but three of the animals that were running loose fell into the hands of the Indians. About twenty miles from the Ford they came upon another train of seven wagons and twelve men that had no stock, all of it having been stampeded and driven off, and they were forced to burn their wagons, and go on foot the balance of the way to California. Later the same season another train was served in the same way, all its stock being taken; but pursuit of the Shoshones was made under the leadership of one — Warner, resulting in a surprisal of the Indians, the killing of some thirty of them, and the recovery of the stock. This put a stop to troubles that season.

In the spring of 1851, Walter Cosser, now living in Douglas County, in this State, left Salt Lake for the purpose of going to California. There were five men accompanying Cosser's party, among whom was the since notorious Bill Hickman, the Danite, or destroying angel of Brigham Young. The five were under the leadership of Hickman; and while they were camped at Stony Point, on the Humboldt River, some Shoshones were standing around, when one of the Danite gang shot and killed a couple of them. Their only reason given for doing it was the pleasure that killing of redskins afforded the murderers. Three or four days later, while upon the same river, Hickman's satellites killed two more Indians and a squaw, and scalped the former. As before, they made no attempt at justifying their acts by accusing their victims of having committed a wrong.

In the fall of the same year (1851) Col. A. Woodard of Sacramento, California, in company with two guards named Oscar Pitzer and John Hawthorn, were carrying the mail from Salt Lake to Sacramento, and camped one night at the scene of Hickman's massacre. That night a mortal tragedy was enacted there among the mountains, by the banks of the Humboldt River; but its silent, passing waters, told no tale. The next traveler over the route found the mangled bodies of three white men at Stony Point, and the newspapers of the Pacific Coast recorded the fact as another outrage on the overland road by savages, and demanded an extermination of the tribe. Where the river came nearest to the rocks a number of willows were growing, and the horsemen, as they approached this place, leveled their rifles at it and rode quietly along, turning in their saddles as they passed, to enable them to continue facing the point of danger. Thus they made their way along by the willows to a more open and safe locality. As they passed beyond rifle

A group of Paiute Indians in Nevada, 1871

Public Library

continued from far left column

Indians and found them to be Pah-Utes, to which I call your attention to the paper marked A. Immediately on receiving this report, and agreeable to the said treaty, I sent Capt. William Weatherlow and Thomas J. Harvey, as Commissioners, to proceed to the Pah-Utes' headquarters, and there inform the chief of this murder and demand redress. Here allow me to call your attention to the paper marked B. It is now pretty well an established fact that the Pah-Utes killed those eight men, one of them being Mr. Peter Lassen. How soon others must fall is not known, for war is now inevitable. We have but few good arms and little ammunition.

Therefore, I would most respectfully call upon you for a company of dragoons to come to our aid at once, as it may save a ruinous war to show them that we have other help besides our own citizens, they knowing our weakness. And if it is not in your power at present to dispatch a company of men here, I do most respectfully demand of you arms and ammunition, with a field-piece to drive them out of their forts. A four or six-pounder is indispensable in fighting the Pah-Utes. We have no Indian Agent to call on, so it is to you we look for assistance. I remain your humble servant,

ISAAC ROOP, Governor of Nevada Territory.

On one of their [Kit Carson & friends] hunting and trapping expeditions in 1847, while camping on Humboldt river, a company of emigrants had several horses run off in the night by the prowling savages. Four of the emigrants went in pursuit of the Indians to recover the stolen stock. When word came to Carson's camp of the loss of the animals, he, with Maxwell, Owen and two Delaware Indians, took up the trail and dashed off to the rescue. And well it was they had taken such a hasty departure, for after a rapid ride of several miles they reached a small valley in the foothills where the savages had entrapped the inexperienced emigrants. Some had pushed on with the stolen animals, while others had remained in ambush until the white men had passed. As soon as they realized their dangerous condition, they entrenched themselves among the rocks and trees as best they could, and being well armed, were making a gallant defense. The Indians, however, were gradually closing in upon them by skulking from one rock and tree to another. Dogs were barking, and women and children shouting when Kit and his followers dashed in with loud, ringing shouts, dealing death with their well-aimed rifles and making a number of "good Indians." The village was soon cleared of the remaining bucks, women and children, the animals recovered and brought back to camp, to the great joy and relief of those dependent upon their teams to pursue their journey.

David Augustus Shaw, 1850

Group of Paiute Indians, probably 1860s

range, however, and lowered their weapons, a number of Indians sprang out from their willow ambush, yelling and gesticulating in impotent rage at the escape of their proposed victims.

WASHOE RAIDS. In the summer of 1852, a man who kept a station on the overland road at a point near the present site of Empire, came up to Eagle Station and informed those stopping there that a band of Washoes on the east side of the river, near that place, had in their possession several American horses that he supposed, of course, they had no right to. It was immediately determined by all to go down and take the animals away from the Indians. The whites, under the leadership of Pearson, a noted Indian fighter, consisted of Frank Hall, now of Carson, his brother, W. L. Hall, of Esmeralda County, the station keeper, and a man named Cady. They found the Washoes with little trouble, but failed to discover the American stock. They found also, that the squaws were taking the unnecessary camp equipage of the band, up the mountain to the east. This looked like business, and when a body of about sixty warriors with their paint on, advanced upon them, matters assumed a decidedly hostile appearance. Pearson, the leader, decided that there were too many to justify risking a fight, and with two of his followers "lit out" Frank Hall and — Cady concluded to await the approach of the enemy and "play the friendly dodge," which they did by distributing their small stock of tobacco among them. Of course the Indians did not object to the gifts, but, after accepting them, ordered the donors to hunt their eyrie at the base of the mountain in the west, and they hunted.

In 1852, the Indians made many raids upon the stock in Carson Valley. In retaliation the whites captured a couple of the tribe and brought them into the Mormon Station as hostages, for a return of the stolen property. One of the captives was a powerful man, dressed in a full buckskin suit, and the other was a mere lad, some sixteen years of age, who dressed as nature had clothed him. Several days passed and nothing was heard from the lost animals; when one morning the larger Indian was let walk out a little way by himself, and he suddenly made a dash for freedom. He scattered his garments as he went, and naked as he was born, bounded like a frightened stag away toward the mountains. The guard, named —————— Terry, had in a careless way leaned his gun against the stockade, and was probably

Paiute Indians at Reno, 1874

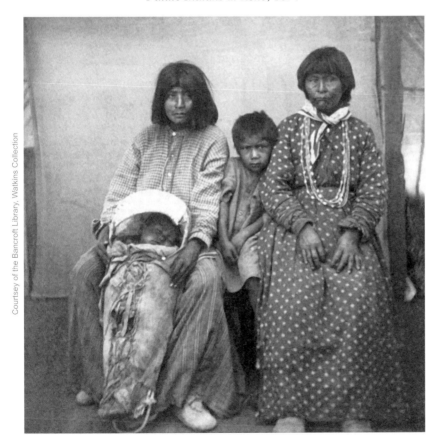

Courtsey of the Bancroft Library, Watkins Collection

The Paiutes drew to council to consider the remissness of their medicine-men. They were sore with grief and afraid for themselves; as a result of the council, one in every campoodie was sentenced to the ancient penalty.

But here at Maverick one third of the campoodie died, and the rest killed the medicine-men.

Mary Hunter Austin, 1903

Dear Mary,

... We now came to a tribe of very hostile Indians, like those we had been with on the Hum-boldt—they are a thieving set; they would come near at nightfall and either steal mules, horses, or cattle, or shoot them with arrows so that they could not be taken along, and then come in and get them after the emigrants are gone. We keep strict guard and save ours. We passed five hundred miles among the robbers; in fact, we are only two days beyond them. Some desperate encounters have been had between them and the whites, when in search of cattle or mules; for they fight well cornered, but run if they can. Yet I have been in the mountains alone by day and by night, have slept alone when the wolves have come howling within two rods of me, and have met with no trouble whatever from either Indi-ans, robbers, or wolves; still, it was a risk.

Alonzo Delano,
1849

Shoshone Indian camp in the Wasatch Mountains, perhaps late 1800s

ten yards away from it when the warrior started; but in a moment he had the formidable rifle in his grasp, and taking a long, deliberate aim, fired. As the whip-like report broke upon the morning stillness the runner leaped high into the air and then fell to the ground; and when they had reached the fallen Washoe, he was dead.

MURDER OF PETER LASSEN. In March, 1859, some prospectors went over from Honey Lake Valley to search for gold in the Black Rock country, in what is now known as Humboldt County. Some of them had been there before, consequently the party separated, four going in advance of the other three. They had an understanding that they were to meet in a cañon on Clapp Creek. where running water is to be found during a portion of the year. The creek is about twenty miles northwest of Black Rock. The second party consisted of Peter Lassen—after whom a peak in the Sierra Nevada Mountains is named—accompanied by —— Clapper and —— Wyatt. They had reached the mouth of the cañon up which the rendezvous had been appointed, as night came on, and camped by a large boulder till morning. At daylight Lassen got up, lit his pipe, sat down and was smoking, when the party was fired on by a concealed foe, and Clapper was killed. Lassen sprang to his feet, rifle in hand, and scanned the surrounding rocks in search of the assailants, but unable to see any, told Wyatt to move their camp equipage to a safer place, while he watched and kept the enemy at bay. The latter had taken one load of their effects away, and was returning for more, when another volley from among the twilight shadows rang out on the morning air; and the brave old hero of many a mountain battle sank down by the rock where he had been standing. As Wyatt came up he said to him, "I am done for at last; take care of yourself;" and, mounting a bare-backed horse, the only survivor, dashed away over the rocks and plains of sand to bear the sad news to the settlements.

INDIAN ACCOUNT OF THE WAR OF 1860. The publishers of this work, desiring the most minute particulars of this most important Indian war of Nevada, in the latter part of 1880 dispatched one of their corps of writers to thoroughly examine the ground and interview all whites and Indians who could be found who had participated in the fatal battle.

It was a strange assemblage, of those old braves, each narrating what he had done, and seen, of that bloody record of 1860. Each Indian would recount his own experience and observation; but when asked concerning anything beyond that, would say: "Me no see 'um mebe ———— tell you 'bout that";

"The California valley cannot grace her annals with a single Indian war bordering on respectability. It can, however, boast a hundred or two of as brutal butchering, on the part of our honest miners and brave pioneers, as any area of equal extent in our republic . . ."

Hubert Howe Bancroft, 1850s

This photo is captioned "California Indian War, 1872." It, and the subsequent photo with the same caption, serves to remind us that it wasn't just Nevada that had Indian wars, they occurred everywhere the white man began to encroach onto Indian lands and threaten their normal lifestyle and culture.

Courtsey of the Library of Congress

Imagine how you'd feel if your nation's culture was distroyed, your family's home and land pillaged, and most of your race was wiped out from the diseases brought by an invader. This Indian was overlooking the Central Pacific Railroad line from the top of Palisades, 430 miles from Sacramento. For him, it was likely just one more gut-wrenching insult against his land and culture.

 The [native] women wear upon the tops of their heads a pyramid of flax and foreign matter, almost a foot in height, and immense rings in their ears made of sea shells. This custom of wearing the hair in this manner is said to be encouraged by the priests of the hoodlum men, and which is called society, because it is known to affect and suppress the feminine brain; it is generally supposed that women with well developed, healthy brains are dangerous to society.

Caroline M. Nichols Churchill, 1874

and the party designated would be sent for, if not present, and the story would go on. On the third day we rode over the battlefield and trail from Pyramid Lake to Wadsworth, a distance of eighteen miles, accompanied by some of them. As we came to a place where a white man had been killed, or some special event worthy of note had transpired, they would stop, and, in their peculiarly slow, dreamy way, tell the event, or describe the death struggle. Their speech was accompanied by gesticulations, and movements of the body, conveying to the looker-on a knowledge of what had transpired there in all its tragic detail before the interpreter had opened his lips. In this manner those events, that before had remained a secret between the slayer and his dead, were revealed.

In the latter part of April, 1860, the Pah-Utes congregated at Pyramid Lake from all over the extensive territory, for the purpose of holding a council. The object of the gathering was to decide what they should do, in view of the fact that the whites were rapidly encroaching upon their lands; killing their game; and cutting down their orchards. [Thus referring to the pine-nut trees.] By the first of May they were nearly all in at the rendezvous.

Numaga's Effort for Peace. Among all that assemblage of the Pah-Ute tribes there was one, and one only, among the chiefs, with sufficient sagacity to foresee the evils that would result to his people from war; one only who at the same time possessed the courage to throw his influence in opposition to their will, and declare for peace. The name of that warrior was Numaga; and he was called by the whites Young Winnemucca, the war chief. The word Numaga means the giver of food, the name indicating the disposition of its owner as being that of a generous man. Numaga was not, as the whites always supposed, the war chief of the Pah-Utes. There was but one general chief, and that was Poito, at Pyramid Lake.

Numaga was the chosen leader only of that branch of the tribe living upon the reservation, having no authority, and claiming none, in any other locality. Neither was he a relative of Poito, and the two were always unfriendly.

Numaga was an Indian statesman who possessed intellect, eloquence, and courage combined. He had been among the whites in California, and could speak the English language; consequently, appreciated the superiority of the race with whom his people would make war. His power, outside of his own band,

California Indian War, 1872

Courtesy of the Library of Congress

"What about the white men that you rode among in the narrow pass?" we inquired.

"White men," said our informant "all cry a heap; got no gun, throw um away; got no revolver, throw um away too; no want to fight any more now; all big scare just like cattle; run, run, cry, cry, heap cry, same as papoose; no want Injun to kill um any more; that's all."

Washoe Indians, the chief's family, 1860s

"Every train that has been attacked acknowledge that they were perfectly unprepared for defense. The Indians watch the trains from the hills, and if they see a train well-armed and watchful, they do not molest it. I have seen many trains on the road during the summer which had plenty of arms, but they were carried in the wagons, and in many cases without being loaded. The emigrants would laugh at me when I told them of the necessity of always having their arms ready for instant use."

Major Lynde, in his report to General Johnston, of October 24, 1859

was that only of a superior mind, working, under the control of an absorbing wish, to better the condition of his race. They knew he was capable, they believed him to be sincere, and it resulted in giving him an influence more potent throughout the tribe than Poito's commands; consequently, the whites came to look upon him as the war chief, and he would have attained that position had he outlived Old Winnemucca alias Poito.

Such was the man who threw himself with all his power into the council, to try, if possible, to stem the tide that had set for war. He rode from camp to camp, from family to family, friend to friend, reasoning, counseling and beseeching them not to precipitate a war, and bring destruction upon themselves. On every side he was met with a calm, respectful silence, that told as plainly as words could have done it, that all were against him. Then he went off by himself, and, lying, down, with his face to the ground would speak to no one. Without food, or drink, or motion, he lay there as one dead. The day passed and the night, another day and night, and the third found him as had the first, a motionless and silent mourner, brooding over the calamity that he saw threatening his people. This began to effect a reaction among the masses of the Pah-Utes, and the chief, seeing it, came to him and said: "Your skin is red, but your heart is white; go away and live among the pale-faces." Others came and said: "Get up or we will kill you;" and then he replied: "Do it if you wish, for I don't care to live."

At length the council met. Chief after chief rose and recounted the wrongs of his band and demanded war. After all had spoken, then Numaga, looking like the ghost of a dead Indian, walked into the circle, and for an hour poured forth such a torrent of eloquence as these warriors had never listened to before:—

"You would make war upon the whites," he said; " I ask you to pause and reflect. The white men are like the stars over your heads. You have wrongs, great wrongs, that rise up like those mountains before you; but can you, from the mountain tops, reach and blot out those stars? Your enemies are like the sands in the bed of your rivers; when taken away they only give place for more to come and settle there. Could you defeat the whites in Nevada, from over the mountains in California would come to help them an army of white men that would cover your country like a blanket. What hope is there for the

An old Indian woman with baskets, identified on mount as Teha, 120 years old. Photo 1905

Courtesy of the Library of Congress

About the first of August [1856] a band of Shoshone Indians camped near where I herded sheep. Some of them could talk Gosiute language which, thanks to my little Indian brother, I could speak, too. They seemed to take quite a fancy to me and would be with me every chance they could get. They said they liked to hear me talk Indian, for they never heard a white boy talk as well as I did. Finally the Indian came one day as he had been doing and let me ride quite a while, and when I got tired and gave him back the horse, he asked me if I did not want to keep him. I told him that I would sooner have the pony than anything I ever saw. He said I could have the horse, and I could ride him all the time, if I would go away with him. I said I was afraid to go. He said he would take good care of me and would give me bows and arrows, and all the good buckskin clothes I wanted, if I would go. I asked him what they had to eat. He said they had all kinds of meat and berries and fish, sage chickens, duck, geese, and rabbits. I thought this surely beat herding sheep and living on greens and lumpy-diet, so I told him I would think the matter over. When he came the next day I told him that I had decided to go with him. . . . Now, my dear friends, how many little boys of my age, lacking only a few months of being twelve years old, would run away from home and friends and off with a tribe of wild Indians? . . . I went with them and for two years I never saw a white man.

Elijah Nicholas Wilson, 1856

Description by photographer Edward S. Curtis: The coiled baskets produced by this woman have not been equalled by any Indian now living. About ninety years old, Datsolali (Dat So La Lee) appears to be in the early sixties. She has the pride of a master in his craft, and a goodly endowment of artistic temperament. Datsolali (said to mean "big hips") is a nickname. Her proper name is Tabuta.

During our trip down the Humboldt we have not been molested though other trains have suffered from the depredations of the indians. In the neighborhood of Stony Point about half way down the river they were particularly trouble some. Cattle were driven off and sometimes killed with arrows and left along the road, others were taken away into the hills. But few of us saw any indians except at a distance. I did not see any myself.

William Henry Hart, 1852.

Pah-Ute? From where is to come your guns, your powder, your lead, your dried meats to live upon, and hay to feed your ponies with while you carry on this war. Your enemies have all of these things, more than they can use. They will come like the sand in a whirlwind and drive you from your homes. You will be forced among the barren rocks of the north, where your ponies will die; where you will see the women and old men starve, and listen to the cries of your children for food. I love my people; let them live; and when their spirits shall be called to the Great Camp in the southern sky, let their bones rest where their fathers were buried."

As Numaga was thus making a last desperate effort to change the action of the chiefs, and was sending home conviction of its folly to their understanding, an Indian, upon a foam-flecked pony, dashed up to the council ground, and the speaker paused. The new-comer walked into the circle; and, pointing to the southeast, said: "Moguannoga, last night, with nine braves, burned Williams' station, on the Carson River, and killed four whites." Then Numaga, with a sad look in the direction that the warrior had pointed, replied: "There is no longer any use for counsel; we must prepare for war, for the soldiers will now come here to fight us."

The steamer "Governor Stanford" delivers mail to Tahoe City in 1873.

Chapter 7: Placer and Nevada Counties - Tahoe City, Truckee and vicinity

Tahoe City is justly conceded by all to be the best point of observation for a general panoramic view of the lake, as from here almost every location of interest is within the range of vision aided by a field-glass of ordinary power. Glenbrook, fourteen miles across the water, is distinctly seen. During winter the snow often falls at Tahoe City to the depth of five or six feet, and in summer the climate there is called the coolest of any place upon the lake. The water of the lake is wondrously clear and blue, so that when in repose fish and other objects can be readily discerned at a depth of thirty or forty feet. It is also very cold, but has the peculiarity of never freezing in the winter. The deepest soundings ever made were 2,800 feet.

Following the lake shore from Tahoe City, the Island House is approached in a distance of a couple of miles, over a fine stretch of country; thence Observatory Point, a sharp

. . . the Indians of the present day never cross the lake, affirming the belief that an evil spirit would draw them to the bottom were they to make the attempt.

Many moun-
tain valleys
of small size
are found in the eastern
portion of Placer County, that
are among the best in the
world for summer pastur-
age for horned cattle, and
for dairying purposes, the
herbage being sweet, and not
causing distasteful flavor
to dairy products, while the
cold, pure water insures
cleanliness and solidity to the
article. Nearly all of these
are occupied for this business,
and a great deal of butter is
made, which, as a rule, finds
ready market without leaving
the mountains—at the tourists' resorts, the logging and
wood-chopping camps, lum-
bering mills—and it is from
this source that the well-to-
do resident of the Silver State
usually secures his annual
supply.

The Custom House, Tahoe City

prominence running into the lake at the lower end of Carnelian Bay; following up the beach, where are found many smoothly-worn and variagated silicious pebbles, the rocky point on the north is passed, and the shore of Agate Bay greets the traveler. Not far from here a small creek enters the lake, about the mouth of which is some pretty meadow land. Griffin's saw-mill is on this stream. East of this a few miles are the Hot Springs, near the State line, now the property of Sisson, Wallace & Co.

The lake is twenty-two miles long and twelve and a half wide, and is fed by the waters of more than thirty streams of various sizes, which have their sources in the surrounding snow-clad hills, and are ever pouring their volume into it—sometimes in gently flowing brooklets; at others in leaping, laughing, beautiful cascades, and again in fierce and angry torrents.

Tahoe City which was first laid out in 1863, by a party of men who, having congregated during that summer and fall at the new diggings in Squaw Valley District, in anticipation of the commercial importance of the place, and in view of its commanding position, located "city" lots, each proprietor digging a trench around his plat. It lies on a gently sloping plateau, at an elevation of about fifty feet above the water of the lake, to mark the boundaries. The following year a wharf was begun by John Chesronn, which was afterwards purchased by J. O. Forbes, Jr., and J. B. Campbell, who completed it. It extends into the lake some 200 yards, and is constructed upon cribs built of strong timbers and anchored with rock—the bottom found in the lake here being too hard to drive piles to sufficient depth in. About 300 feet from the shore, upon the wharf, is situated the "Custom House," a building used as a saloon and post-office, now owned by J. B. Campbell. Beyond the wharf, some 200 feet into the lake, one strong pier has been sunk, where steamers are moored in bad weather, when too rough to lay up at the wharf. The first public house built there was the Tahoe House, by William Pomin, who is yet the owner and proprietor. He also built a brewery there. Later, as the place began to be visited by pleasure-seekers, as steamers began to ply the lake, and a wagon road was constructed from Truckee, after the completion to that point of the Central Pacific Railroad, the want of more extended accommodations was felt, and another hotel was built. This, after passing through numerous managements, has at length merged into one of California's palatial hostelries under the ownership of A. J. Bayley, and is now enduringly established, with a world-wide reputation, as the Grand Central Hotel.

The marine interests of Placer County were first noticed by the Assessor in 1866, when he, that year, listed upon the assessment roll for taxation two schooners which were then plying upon the lake. The lake portion of Placer did not become populated as early as some other parts, owing to its isolation

Tahoe City, 1899

Steve Crandell Collection

...let us simply note the thrill of awe and wonder with which we gazed up the walls of the Blue Cañon, one thousand feet of sheer precipice, while far below winds a narrow ribbon of blue water curving to the curve of the foothills, and sweeping around their craggy feet, avoiding the jagged points, and lending grace and beauty to the stern and rugged scene in a manner altogether feminine. By the way, how true is that instinct which makes every one call a river she, and a mountain he!

Miriam Squier, 1877

Jorge Ballen, or, as he is commonly called, Greek George, built a house in Little American Valley, and kept store as well as public house for the accommodation of travelers. He and his wife concluded at one time to remain there all winter, and brave the rigors of the climate. The one hibernation there, however sufficed, as the snow fell early and remained late, and most of the time was twenty feet deep. Mrs. Ballen did not leave the house for a period of five months, and then made her first appearance from that long imprisonment in the month of April, after her husband had shoveled a trail from the door of the house to the roof, to which she ascended, and there, upon a blanket spread for the occasion stood for awhile to bask in the sunlight from which she had been so long deprived—the snow at that time being, by actual measurement, just five feet deeper than the apex of the roof.

Gents and ladies enjoying some lakeside leisure time at Lake Tahoe

from the direct routes of wagon travel. About the earliest permanent settlements there, of which there is any record, were those at the mouth of McKinney's Creek, at Ward's Creek, and at the outlet, now Tahoe City, in 1861 and '62. In the winter of 1861, a man who attempted to pass the winter on the lake, near the outlet, was frozen to death. William Ferguson and Ward Rust built a cabin on the lake at the mouth of Ward's Creek, in the summer of 1862, having gone there from Volcanoville, El Dorado County. John W. McKinney and Thomas Wren located a hay ranch on the summit, near the county line, in 1861, but McKinney, in 1862, went to the lake shore, near the creek now bearing his name, and located there, for the purpose of hunting, fishing, and trapping, where he has ever since remained, and has established quite a noted place of resort known as

THE HUNTER'S HOME, Which is patronized extensively by people from the State of Nevada, as well as by tourists. He has erected, for the accommodation of guests, besides the main building in which is the dining-hall, comfortable cottages, to the number of twenty-five or more, bordering the clean, pebbly beach, just far enough back from the lake to be away from the reach of high water. A good wharf extends into the lake some two hundred feet to water deep enough to admit of steamer landing, on which is a good saloon building 22x32 feet in dimensions, and two stories high. All of the steamers running on the lake stop semi-daily at McKinney's Landing for passengers and mail, and the old pioneer always treats his guests well. He always keeps a number of sail and row-boats for yachting or fishing parties. Among the former is the Transit, the crack yacht of the lake, and the fastest sailer.

Going northerly along the lake shore, from the Hunter's Retreat a mile distant, and the mouth of Madden's Creek is reached; two miles further to Blackwood, where there are nice picnic grounds, with a large floor forty feet square laid for dancing, and a wharf where steamers land. Thomas McConnell, of

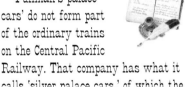

'Pullman's palace cars' do not form part of the ordinary trains on the Central Pacific Railway. That company has what it calls 'silver palace cars,' of which the name is the best part. They are very inferior when compared with those of the Pullman Company. Besides, the system of management is far less perfect.

William Fraser Rae, 1869

The elegant Grand Central Hotel built at Tahoe City in 1871 had beautiful walnut furniture, imported carpets, and an ornate kitchen range. It surpassed the Glenbrook House as Tahoe's most fashionable resort but was then surpassed itself by Elias J. "Lucky" Baldwin's Tallac House further down the beach.

Courtsey of the Library of Congress

Once the railroad crossed over the mountains loggers utilized the trains whenever possible. There were frequently "spurs" built into specific areas to utilize rail transportation over the slower teamsters.

PRESCOTT, Arizona Territory, Feb. 20, 1881.

MY DEAR SIRS: What is now called Tahoe Lake I named Lake Bonpland, upon my first crossing the Sierra in 1843-44. I gave to the basin river its name of Humboldt, and to the mountain lake the name of his companion traveler, Bonpland, and so put it in the map of that expedition. Tahoe, I suppose, is the Indian name, and the lake the same, though I have not visited the head of the American since I first crossed the Nevada in '44.

Yours Truly,
J. C. FREMONT.

Sacramento County, owns the land, and has a summer cottage there. Here, too, is the "Wildidle" cottage, belonging to Mrs. Crocker and daughter, of Sacramento, besides another belonging to some resident of the State of Nevada. Other people have bought lots in Blackwood, and will build cottages during the summer of 1882. Two miles further toward the foot of the lake is where Ward's Creek debouches, and here is the "Sunnyside Cottage," belonging to Mrs. Hayes, of Nevada State, with wharf, boat-house, etc. About a mile from the latter-named stream is the site of the saw-mill of Reuben H. Saxton, on the lake shore, which, when running, was propelled by an overshot wheel fifty-four feet in diameter, run with water brought in a ditch from Ward's Creek. All of these streams are resorted to by trout in spawning-time, when thousands of them of large size are taken.

The next point of interest reached is the Truckee River, the outlet of Lake Tahoe. The stream here is about fifty feet wide, with an average depth of five feet, the water flowing with a velocity of five feet a second, and discharging from the lake approximately 518,400,000 gallons of water every twenty-four hours.

LAKE TAHOE'S NAME. The name of this grand lake now appears to be fixed as Tahoe, but over this subject an exciting and acrimonious controversy has more than once been held. The first record of the lake is in Fremont's explorations of 1843-44. January 10, 1844, he discovered and named Pyramid Lake, and a few days thereafter reached the river at the south end, where he had a feast of salmon trout, and he named the stream Salmon Trout River. This now bears the name of Truckee. He was told that the river came from another "lake in the mountains three or four days distant, in a direction a little west of south" On the maps accompanying "Captain Fremont's Narrative," this lake bears the name of "Mountain Lake," and it was so called in California until after 1852, and it is thus referred to in the Placer Herald in discussing the subject of wagon roads. In 1852, the Surveyor-General, looking out a route for a wagon road, gave the name of the then Governor of California to the lake, and it became officially and generally known as Lake Bigler.

SQUAW VALLEY, lying at an altitude of 6,126 feet above the sea. During the time in the history of the State when there were so many emigrant wagon-roads projected, the Placer County road, as contemplated, ran through it.

The *deboucheur* of the stream running through Squaw Valley into the Truckee is about five miles from the lake. This beautiful valley has been utilized by the farmer, its products of hay and vegetables, butter, cheese, eggs, and berries, usually

Among the French immigrants of 1849 were maps in which the mountain lake was given the name of "Bonpland." This name had been given to it by Preuss, the draughtsman accompanying Fremont in 1843-44, and was so published in Europe. This name seemed so appropriate, as in honor of a great traveler and geographer, the famous companion of Baron Von Humboldt, that, when the change was proposed to Tahoe, an effort was made to have the name of Bonpland re-established. This, however, did not obtain.

This Taber photo taken at Blue Canyon probably in the late 1800s certainly illustrates some of the travel problems the early pioneers faced. Even with modern transportation, traversing the Sierras is often a formidable task.

Lots of trees coming down and buildings going up in Truckee in 1869! This view of its wooden buildings looks west along Front Street.

Squaw Valley is the most beautiful valley the eye ever beheld. It is covered with luxuriant grass and the soil is of the most productive nature. The valley is completely surrounded by mountains, with the exception of the east end, at which point a most magnificent stream of water, that flows through the entire length of the valley, empties into Truckee River. There is contained in the valley about 500 acres, of tillable land.

Surveyor Thomas A. Young, in his report to the Surveyor-General

finding a market at the hotels on Lake Tahoe, the saw-mills of the region, and in Nevada.

TRUCKEE. The name Truckee was given to the home of the leaping trout, the beautiful river that receives its waters from lake Tahoe and carries them swiftly through this enchanting valley.

Upon a petition being presented to the Board of Supervisors, signed by 160 legal voters of the town, that body incorporated the town as "The Inhabitants of the town of Truckee," under the provisions of the Act approved April 19, 1856. The town government was to consist of five Trustees, Treasurer, Assessor and Marshal, to be elected January 18, 1879 and hold office until the first Monday in May, 1879, at which time and annually thereafter, the regular town election should be held.

It [the town] is of wood, but is well protected from fire. Two snow plows are kept here, and during the winter see much service. A fire engine and train for use in protecting railroad property, especially the snow sheds, are kept here. There is also a repair shop 30x150 feet in size. Truckee has a fine school house, built in 1871, at an expense of $2,200, the old one having been burned in the fire of July 20, of that year. Smelting works existed here for a time, on account of the cheapness of charcoal, but were removed.

Three hotels open their hospitable doors to the public. The Truckee Hotel, J. F. Moody, proprietor, has always been the headquarters of the railroad. It stands by the track and contains the ticket office and waiting room. It was built immediately after the fire that destroyed Coburn's Station, in 1868, and was then known as Campbell's Hotel. J. F. Moody became the proprietor in 1870. It contains sixty-five rooms with sleeping accommodations for one hundred. Excellent meals are furnished here to travelers and railroad men, and from 150 to 200 sit down to the tables at every meal. Stages for Lakes Tahoe and Donner and for the Sierra Valley leave the hotel, and Harry Hollister, Mr. Moody's accommodating assistant, takes pleasure in giving information about the wonders to be seen, in regard to which no one is more familiar than he.

The railroad winds around the face of the mountains, far above the quiet lake beneath, glimpses of whose mirrorlike waters can be hastily snatched by the traveler, between the long rows of snow sheds, through which the train so securely glides. Lake Tahoe is one of the attractions of Truckee. This beautiful lake, lying among Sierra's crowning peaks, is one of the grandest scenes in California, and is one of the seven wonders of the coast. Thousands of delighted pleasure seekers stroll along its beach and sail over its clear waters during the summer season. Weber and Independence lakes and the Sierra Valley are all

Those who have been accustomed to the monotony and conventionalities of European travel, will find an immense relief and sense of freedom when roaming amongst the Sierras of California; and it cannot but be interesting to visit a state that, less than a quarter of a century since, was hardly known by name, and to-day takes her place as one of the powers of the American Continent.

William Henry Brewer, 1860s

Lumber mill at Rocky Glen on the Truckee River

Courtesy of the Bancroft Library, University of California, Berkeley

Hauling freight from Truckee, 1909. Even with the Central Pacific Railroad trains running regularly over the Sierra for many years, and lots of speculation that the railroad would put teamsters out of business, many persisted into the early 20th century.

""Do many people get killed on this route?" said I to Charlie, as we made a sudden lurch in the dark and bowled along the edge of a fearful precipice. "Nary a kill that I know of. Some of the drivers mashes 'em once in a while, but that's whisky or bad drivin'. Last summer a few stages went over the grade, but nobody was hurt bad—only a few legs'n arms broken. Them was opposition stages. Pioneer stages, as a genr'l thing, travels on the road. Git aeoup!" "Is it possible? Why, I have read horrible stories of the people crushed to death going over these mountains!" "Very likely—they kill 'em quite lively on the Henness route. Git alang, my beauties! Drivers only break their legs a little on this route; that is, some of the opposition boys did it last summer; but our company's very strict; they won't keep drivers, as a genr'l thing, that gets drunk and mashes up stages. Git aeoup, Jake! Git alang, Mack! 'Twon't pay; 'tain't a good investment for man nor beast. A stage is worth more'n two thousand dollars, and legs costs heavy besides. You Jake, git!"

J. Ross Browne, 1871

objects of interest to the traveler, who makes Truckee his base of operations. The Sierra Valley, lying forty miles north of Truckee, and with which it is connected by telegraph and stage line, is one of the tributaries of Truckee, to which it sends wool and butter for shipment. A railroad from Truckee to open up the vast timber interest of this region is among the possibilities of the near future.

Truckee contains three hotels, four grocery stores, three dry goods stores, two general merchandise stores, one clothing store, one drug and stationery store, one variety store, one hardware store and tin shop, one boot and shoe store, one furniture store, two markets, two livery stables, three breweries, one bakery, one carriage paint shop, one photograph gallery, ten saloons, two jewelers, two blacksmith and wagon shops, one tailor, one newspaper, one dentist, two physicians, four attorneys, one bank, one post office, one Wells Fargo & Co. express office, one school house, factories and saw mills as described elsewhere, railroad round house and shops, one church, a number of handsome residences and a great many comfortable and neatly kept cottages. The population [in 1880] is about 2,000 whites and an indefinite number of Chinese, ranging between 500 and 1,000; besides these some fifty Washoe Indians hang around the town.

With a railroad to Sierra Valley and another to Tahoe City, in its possible extension to Carson City, Truckee would reach the object of her desires, to be the central shipping point of this vast lumber region.

Truckee was made the end of one division of the road [Central Pacific Railroad], and a round house and necessary shops were built. The number of stores was increased, three hotels were built, many new residences were erected, several saw mills were in operation in the neighborhood, and the town started at once on the path of prosperity. The railroad round-house was burned March 28, 1869. It was evident that an incendiary's hand had applied the torch, and D. J. Hickey, to whom suspicion pointed strongly, was arrested and indicted for arson. His trial lasted four days and resulted in a disagreement of the jury. The same result followed the second trial, and he was then discharged.

At this time Truckee was the chief town on the railroad between Sacramento and Ogden.

The first [conflagration] one was in January and the second in March; the burned buildings had but scarcely been replaced when the last and most destructive one occurred, July

Snow sheds along the Central Pacific Railroad

Courtesy of the Library of Congress

Dear Mary,

I preferred sleeping near a camp, for this forest swarms with grizzly bears and large wolves and panthers, their tracks being very frequent in the road.

Alonzo Delano, 1849

Front Street, Truckee immediately after one of the fires of 1871 that burned all of Front Street and in one fire, almost the entire town. Some of the buildings being reconstructed are of stone which were considered "fireproof" but were not always so. The photo is looking east along the railroad. At the very right edge of the photo is the Truckee Hotel that did not burn in any of the fires because it was on the other side of the tracks.

December 22d.[1864]— Mr. Stevens, a cattle-dealer from Yolo County, was robbed by a highwayman between Auburn and Yankee Jim's, and relieved of $550 in coin.

20, 1871. A large public meeting was being held, when every heart was thrilled by the sudden cry of fire. A rush was made by the citizens to Derr's saloon, from which flames were issuing. The most frantic efforts of the desperate people were unavailing to stay the progress of the flames, although the women added their exertions to those of the men. All the business portion of the town except three brick buildings was burned. The railroad property was saved as well.

In all sixty-eight buildings were destroyed and sixty-three families rendered homeless.

It was soon ascertained that the fire was of incendiary origin. Mrs. Derr had had some trouble with her husband, and on the night of the fire he was to return to Truckee from San Francisco. She declared that he should never set foot in the house again, and so set fire to the establishment. As soon as these facts became known the excitement was intense and a determination to lynch her was made by many. She was, however, arrested and E. H. Gaylord was engaged to defend her.

The next experience with fire was the destruction of Chinatown and a few adjacent buildings, about three o'clock on the morning of May 29, 1875. Chinatown, then situated in the heart of the place, and just across a narrow street from the row of business buildings on Front street, had always been a menace to the town. A lot of dry, closely packed wooden shanties,

among which a fire had only to be started to become uncontrol-
lable, and insure the almost certain destruction of the town, it
is no wonder that the citizens watched them with eye. At the
time mentioned a fire broke out here, and threatened to become
a general conflagration. The fire engine, Sampson, was soon
at work, the fire train came rushing down from Summit Station, having been telegraphed for. These with the assistance of
a hose, attached to a hydrant on Second street, and many pails
of water, succeeded in quenching the flames, after the whole
of Chinatown was consumed. Besides this the Virginia saloon,
Cruther's cabinet shop, Grozen & Stoll's stable and Paschen &
Kerby's market were burned. The total loss was about $50,000,
chiefly by the Chinese. An effort was made to prevent the re-
building of the Chinese quarters, but without success.

The property owners soon had occasion to regret that their
fire organization had not been maintained, for on November 6,
1875, the planing mill of Elle Ellen caught fire and in twenty
minutes was destroyed. The fire spread to some tenement houses close by and threatened to reach the business portion of the
town. Some pieces of hose were procured and with these and
buckets of water the flames were subdued, after burning two of

Central Pacific Railroad's third crossing of the Truckee River, late 1860s

January 11th [1859].—The
stage between Forest Hill
and Todd's Valley was
stopped by eight men and the
express box, containing 100
ounces of gold, taken, Several
shots were fired. The robbers
escaped.

Courtesy of the Library of Congress

Inside one of the many snow sheds along the Central Pacific Railroad route across the Sierra, 1860s or 70s

 Once more in train, and, during the next fourteen miles, we ascend nearly twelve hundred feet. Two immensely powerful engines perform the arduous task, until at length we reach Summit Station, seven thousand and seventeen feet above the sea, and only two hundred and forty miles from San Francisco. Here we pass through a railroad construction peculiar to the Central Pacific line, I mean the snow-sheds. Let the reader picture to himself a long gallery composed of immensely strong uprights of timber and great joists of pine wood, the whole arched Gothic fashion, with here and there small openings, through which a glorious panorama is seen for an instant as the train roars its way along.

J.G. Player-Frowd, 1872.

the cottages. Had it not been for the fact that the air was calm and that a drizzling shower of rain aided them, much loss would probably have occurred; as it was the loss was about $17,000. The last fire of any consequence occurred on March 12, 1878, which destroyed the block on Bridge and Church streets, of which the American Hotel was the principal building. The loss was estimated at $20,000. The Truckee Lumber Co. had organized a fire company among its employees for the protection of its property, and these rendered good service at this time, as did also the Washoe Engine Co., No.1, that had been organized in 1877. Their steamer was bought in Virginia City, and their bell which gives the alarm of fire is the same one used by the San Francisco Vigilance Committee in 1856.

The citizens who had resorted to every peaceable means to induce the Chinese to vacate their quarters in the heart of the town, and being thoroughly convinced that their presence was a constant menace, on account of the danger from fire, finally resolved to abate the nuisance. A body of four or five hundred of them assembled in the Chinese quarters, on November 18, 1878, pulled down and totally destroyed Chinatown, giving the denizens notice to leave the town within one week. Beyond the tearing down of the buildings no violence was offered, and no serious disturbance occurred. Within a month from that time a new Chinatown sprang up on the south side of the river and without the city limits. As an instance of the customs of the Chinese which are repulsive to our ideas, the following is interesting. The right of property in women is recognized by them and often defended even against our legal authorities. Ah Quee, of North San Juan, owned a Mongolian maiden named Sin Moy, who was kidnapped by a countryman and brought to Truckee. She brought with her some trinkets, and Ah Quee procured a warrant for her arrest for larceny, simply as a means of obtaining possession of her again. December 17, 1872, the warrant was placed in the hands of Constable Cross, who with a posse of four or five went to the Chinese quarters and attempted to make the arrest. All Chinatown arose in arms to repel the invaders, and a lively conflict ensued, during which some forty shots were fired. The officers secured their prisoner and retired from the field without harm to themselves. Not so with Ah Quee, for he and another Chinaman were seriously wounded, and several Mongolians received slight injuries. An attempted abduction in the evening of January 3, 1874, resulted in a riot and the wounding of half a dozen of the participating Chinamen.

Fallen Leaf Lake, 1860s

Courtesy of the Library of Congress

Up the Sierra at a height of 7017 feet, where the snow lay sixty feet deep one winter while the road was building, and where they actually dug tunnels through the snow and ice to work on the road-bed; down from the summit around cliffs, along the edge of precipices, through miles of snow-sheds, through tunnels and deep rock-cuts, across chasms where you shudder as you look down into the rushing torrent far below; and all this, until you reach the plain of the Sacramento, through a country even yet almost uninhabited, believed ten years ago to be uninhabitable, presenting at every step the most tremendous difficulties to the engineer as well as to the capitalist.

The story of the building of the Central Pacific Railroad is one of the most remarkable examples of the dauntless spirit of American enterprise.

Charles Nordhoff, 1872

Pack train at Dutch Flat, 1860s. Notice the distinct absence of trees. Most forests were harvested to build towns, shore up mines, and fulfill the needs of a rapidly growing population.

On one occasion the express train entered the wooden aisles; on arriving near the end the driver saw that they were on fire. To check the train would be to risk its stopping in the flames. He saw that the fire had only just commenced; he clapped on all the steam he possibly could; he and the stoker wrapped their blankets around their heads, and they dashed through the blinding smoke and flame.

J.G. Player-Frowd, 1872

LAKE PASS (DUTCH FLAT) WAGON ROAD has figured extensively in newspaper and political controversy in connection with the construction and progress of the Central Pacific Railroad.

The discovery of the silver mines of Washoe in 1859 gave a great impulse to travel over the mountains, and every county in which there was a practicable pass was anxious to have a road running through it. In answer to this desire the Legislature in 1860 passed a bill giving the State's portion of Foreign Miner's License and Poll Tax for the years 1860 and 1861 to the counties of Tuolumne, Calavaras, Amador, El Dorado, Placer, Sierra and Plumas for the purpose of enabling them to build roads over the Sierra Nevada. The State's portion of these moneys in the year ending June 30, 1859, in the county of Placer amounted to $17,210.76, and should the same rate continue during the two years the aggregate would be $34,421.52 for this county alone. The people of the counties mentioned were elated by the passage of this bill, which would build in each a good stage road over the Sierra; but their hopes were blasted by the veto of Governor Downey, who declared the bill preposterous, and that the withdrawal of such large amounts from the annual revenue would bankrupt the State.

This scheme so condemned by the Governor was not such a wrongful robbery of the State Treasury as it would seem. The amount appropriated, or to be diverted, was derived chiefly from the Foreign Miners' License Tax—a license of $4.00 a month for working in the gold mines—collected almost entirely in the counties included in the bill, and from the Chinese miners only. The agricultural or "cow counties," were subject to no such tax, but persisted in the "mining counties " paying it into the State Treasury. The law authorizing the collection of the tax was shortly afterwards declared inconsistent with the "Civil Rights Bill" and with United States treaties, and the deprivation of the fund did not bankrupt the State.

The vetoing of this bill forbade the construction of free roads over the Sierra, and several toll-roads were the result, yielding large revenues to their owners. The Lake Pass Turn-pike Company was organized at Dutch Flat, March 21, 1861, for the purpose of constructing a turnpike from that place to Steamboat Springs, in the Territory of Nevada. The treasurer of the company reported having received the sum of $7,500 in cash, being ten per cent, of the capital stock. A contract was let to S. G. Elliott for the construction of the road, for the sum of $66,000, that being the lowest satisfactory bid. The Placer Herald congratulated the people of Dutch Flat upon such a bright promise for their place, saying, "Dutch Flat is now second to no town in the county in population and business, is only thirty-three miles from the summit, and a portion of the distance is a good natural road. From Sacramento to Dutch Flat an ordinary eight-mule team will easily haul 8,000 pounds of freight. By way of Dutch Flat will not only be the great

Truckee River area, date unknown

Courtesy of the Bancroft Library, University of California, Berkeley

"On this night they ate the last flesh of their deceased companions. One of the company then proposed that they should kill the two Indian boys, Lewis and Salvadore, who, it will be remembered, met them with Mr. Stanton, with provisions for their relief; Mr. Eddy remonstrated, but finding that the deed was resolved upon, he determined to prevent it by whatever means God and nature might enable him to use. Desiring, however, to avoid extremities, if possible, he secretly informed Lewis of the fate that awaited him and his companion, and concluded by advising him to fly. The expression of the face of Lewis, never can be forgotten; he did not utter one word in reply, but stood in mute astonishment. In about two minutes his features settled into Indian sullenness, and he turned away to fly from the scene of danger."

Jesse "J" Quinn Thornton, 1848

Three ladies boating, Donner Lake, Mirror Cove, 1872.
Picture taken from the west end looking east.

Dear Mary [Keyes],

". . . we was a traveling up the truckee river we met a man and a Indians that we had sent on for provisions to Suter Fort thay had met pa, not fur from Suters Fort he looked very bad he had not ate but 3 times in 7 days and the days without any thing his horse was not abel to carrie him thay give him a horse and he went on so we cashed some more of our things all but what we could pack on one mule and we started Martha and James road behind the two Indians . . . ."

Virginia Reed, age 13, member of the Donner Party

May 16, 1847

wagon route, but the railroad that must be built not many years hence must follow the same. All success, then, and speed to the Dutch Flat Wagon Road."

But the summer of 1861 passed, and the road was not constructed. The Dutch Flat inquirer of October 10th, says: "We learn that responsible parties will commence work soon on the wagon road across the Sierra. Parties who have passed over the route in light wagons and on horseback represent it as perfectly practicable. Freight teams will be able, when this road is made, to make the trip from Washoe to Auburn in four days."

On the 19th of October of that year Leland Stanford, Governor elect, C. P. Huntington and Charles Crocker, of Sacramento, and Dr. D. W. Strong, of Dutch Flat, left the latter place on a tour of inspection of the route proposed by Judah for the railroad and wagon road. Shortly after the return of these parties the

DUTCH FLAT & DONNER LAKE WAGON ROAD COMPANY was formed, with a capital of $100,000. This company was composed of the same parties who were at that time attempting to make headway in the organization of the Central Pacific Railroad Company. Work was begun on the wagon road in the fall of 1862, and a few miles constructed. In June, 1863 a large force was at work, numbering nearly 500 men, but even with this force the road was not completed when the snow in November drove the laborers from the work.

The road was open for travel early in June, 1864, and it was then said to be the best mountain road in the State. The California Stage Company commenced running over the road on the 16th of July, from the railroad at Clipper Gap to Virginia City, making the trip from Sacramento through in sixteen hours. As the railroad progressed and made stations at various points the stages and forwarding houses also moved on and made connection at the terminus. The railroad company thus forced the stages and freight wagons over their own road, which aroused the suspicion that the railroad was only a feeder to the wagon road. Thus it received the epithet of "Dutch Flat Swindle" from the enemies of the company, which it bore until the railroad had so far progressed as to prove that it really meant to build a great trans-continental road instead of the comparatively small affair for local business. When the railroad had reached Colfax, in 1865, it commanded the greater part of the freight and passenger business between California and Nevada, which was very large, and the revenue to the company was in proportion.

PACIFIC TURNPIKE. "All roads lead to Rome" was said of one historic period, but in the early years of the seventh decade of our century all roads led to Washoe, and among the number was the Pacific Turnpike, or Culbertson's road. The construction of this was undertaken in May, 1863. In June there were 125 men at work and an advertisement in the paper for 300 more. The road led from Dutch Flat, via Bear Valley, Bowman's

This is a Pacific Wood & Lumber locomotive. Note its small size. It was built originally for use in the Sutro Tunnel but instead was sold to this logging company and put to good use in moving massive logs to the mill.

Courtesy of the Truckee-Donner Historical Society

Dear Mary

"well we thought we would try it so we started and thay started again with thare wagons the snow was then up to the mules side the farther we went up the deeper the snow got so the wagons could not go so thay packed thare oxens and started with us carring a child a piece and driving the oxens inn snow up to thare wast the mule Martha and the Indian was on was the best one so thay went and broak the road and that indian was the Pilot so we went on that way 2 miles and the mules kept faling down on the snow head formost and the Indian said thay he could not find the road stoped and let the Indian and man go on to hunt the road thay went on and found the road to the top of the mountain and came back and said they thought we could git over if it did not snow any more we the Weman were all so tirder caring there Children that thay could not go over that night so we made a fire and got something to eat & ma spred down a bufalo robe & we all laid down on it & spred somthing over us & we all laid down on it & it snowed one foot on top of the bed"

Virginia Reed, age 13, member of the Donner Party

November 2, 1846

Crystal Lake House along the Central Pacific Railroad, late 1860s.

"On the twentieth the sun rose clear and beautiful, and cheered by its sparkling rays, they pursued their weary way. From the first day, Mr. Stanton, it appears, could not keep up with them, but had always reached their camp by the time they got their fire built, and preparations made for passing the night. This day they had travelled eight miles, and encamped early; and as the shades of evening gathered round them, many an anxious glance was cast back through the deepening gloom for Stanton; but he came not."

John Sinclair, while in the area of Crystal Lake, December 20, 1846.

Ranch, Henness Pass, Webber's Lake, Sardine Valley, and Dog Valley, to the Truckee River near Verdi, a great deal of the route being in Nevada County. This was six miles shorter than the Dutch Flat and Donner Lake Road, and the grade and road bed was claimed as making it one of the best of the many excellent turnpikes crossing the Sierra. These were completed and opened for travel in May, 1864, and for several years a large amount of the transmountain business passed over them.

PLACER COUNTY AND WASHOE TURNPIKE. The failure of the public to improve the Placer County Emigrant Road left the opportunity open to private enterprise. The discovery of silver in the Comstock vein in 1859, and the rising excitement infused great enterprise among road-builders, and every route possible crossing the Sierra Nevada was sought for the purpose of constructing toll-roads to the land of silver, or "Washoe," as it was then universally called. The route through Placer County via Yankee Jim's and Squaw Valley was known to be practicable, and on the 11th of February, 1860, a company was organized at Forest Hill with a capital stated at 150,000, under the name of the "Placer County and Washoe Turnpike Company," to construct and maintain a road over this route. William N. Leet, an enterprising citizen of Michigan Bluff, was President of the company. The project, however, was never carried to a successful conclusion.

Bibliography

BOOKS, JOURNALS AND EXPERTS

Sioli, Paolo *Historical Souvenir of El Dorado County California*, 1883

Thompson & West, *History of Nevada County*, 1880

Thompson & West, *History of Nevada*, 1881

Thompson & West, *History of Placer County California*, 1882

INDIVIDUALS.

AJAX, WILLIAM, 1832-1899, was a Mormon pioneer. Ajax settled in Utah in 1862 and ran a prosperous store in Rush Valley, Utah. He was 30 when he wrote the journal from which his quotes are taken.

AUSTIN, MARY HUNTER, 1868-1934, moved with her family from Illinois to the desert on the edge of the San Joaquin Valley in 1888. In the next fifteen years she moved from one desert community to another, working on her sketches of desert and Indian life. Spending the last years of her life in Santa Fe, Austin remained a lifelong defender of Native Americans and was recognized as an expert in Native American poetry. *The Land of Little Rain* (1903), Austin's first book, focuses on the arid and semi-arid regions of California between the High Sierras south of Yosemite: the Ceriso, Death Valley, the Mojave Desert; and towns such as Jimville, Kearsarge, and Las Uvas. She writes of the region's climate, plants, and animals and of its people: the Ute, Paiute, Mojave, and Shoshone tribes; European-American gold prospectors and borax miners; and descendants of Hispanic settlers.

AVERY, BENJAMIN PARKE (1828-1875) was a New York journalist who emigrated to California. He became part owner of the *Marysville Appeal* in the 1850s and later published a newspaper in San Francisco and served as state printer. Californian pictures in prose and verse (1878) contains his "word-sketches," which are largely confined to California scenery, although some picture Native Americans and miners whom he knew when he prospected on the Trinity River in 1850 as well as the city of San Francisco.

BANCROFT, HUBERT HOWE, 1832–1918, American publisher and historian, b. Granville, Ohio. Bancroft began his career as a bookseller in San Francisco in 1852. Soon he had his own firm, the largest book and stationery business west of Chicago. He also developed a passion for collecting materials on the western regions of North and South America, from Alaska to Patagonia. After toying with the idea of compiling an encyclopedia, he settled on the publication of a prodigious history (39 vol., 1874–90), reissued (1882–90) as *The Works of Hubert Howe Bancroft*. The Works cover the history and to some extent the anthropology of Central America, Mexico, and the Far West of the United States. The first five volumes concern the native races, the next 28 the Pacific states, and the last six are essays. Literary Industries, the 39th volume, contains autobiographical material and an account of Bancroft's historical method. About a dozen assistants—out of hundreds Bancroft had tried out in his "history factory"—did the actual writing of the Works; Bancroft personally wrote very little. Because his assistants were not given credit lines and because of Bancroft's rather unethical business practices, Bancroft and the Works were at first severely attacked. However, his enormous contribution soon received just recognition. When Bancroft presented his library to the University of California (1905) it contained about 60,000 items, including rare manuscripts, maps, books, pamphlets, transcripts of archives made by his staff, and personal narratives of early pioneers as recorded by his reporters. Known as the Bancroft Library, the collection is an outstanding repository of the history of the West.

BREWER, WILLIAM HENRY (1828-1910) was a professor of chemistry at Washington College in Pennsylvania when he joined the staff of California's first State Geologist, Josiah Dwight Whitney, 1860-1864. On returning east, Brewer became Professor of Agriculture at Yale, a post he held for nearly forty years. *Up and Down California* (1930) collects Brewer's letters and journal entries recording his work with Whitney's geological survey of California, chronicling not merely the survey's scientific work but the social, agricultural, and economic life of the state from south to north as the survey's men passed along. New Haven, Yale University Press; London, H. Milford, Oxford university press, 1930.

BROWNE, J. ROSS, pioneer and author of Crusoe's Island (1864), and Adventures in the Apache Country (1871).

CARTER, HENRY was working at the General Land Office in Washington D.C. and in 1869 relocated his family to California.

CHURCHILL, CAROLINE M. NICHOLS, (b. 1833) moved from Chicago to California in 1870. *Little Sheaves* (1874) recounts her experiences in the West, with special attention to San José, Santa Cruz, Santa Clara, San Francisco, Gilroy, Petaluma, Santa Rosa, Healdsburg, and Los Angeles; and Reno, Carson City, and Virginia City, Nevada.

CLEMENS, SAMUEL LANGHORNE (1835-1910), better known as "Mark Twain," left Missouri in 1861 to work with his brother, the newly appointed Secretary of the Nevada Territory. Once settled in Nevada, Clemens fell victim to gold fever and went to the Humboldt mines. When prospecting lost its attractions, Clemens found work as a reporter in Virginia City. In 1864, Clemens moved to California and worked as a reporter in San Francisco. It was there that he began to establish a nationwide reputation as a humorist. *Roughing It* (1891), first published in 1872, is his account of his adventures in the Far West. He devotes twenty chapters to the overland journey by boat and stagecoach to Carson City, including several chapters on the Mormons. Next come chronicles of mining life and local politics and crime in Virginia City and San Francisco and even a junket to the Hawaiian Islands. The book closes with his return to San Francisco and his introduction to the lecture circuit. Regarding some of the clips from the *Territorial Enterprise*, Twain writes for a monthly magazine, *The Galaxy*: "...I certainly did not desire to deceive anybody. I had not the remotest desire to play upon any one's confidence with a practical joke, for he is a pitiful creature indeed who will degrade the dignity of his humanity to the contriving of the witless inventions that go by that name. I purposely wrote the thing as absurdly and as extravagantly as it could be written, in order to be sure and not mislead..."

COBBEY, JOHN FURMES, 1827-1854, was a native of Illinois and gold prospector who died at the age of twenty-seven. Quotes are taken from his handwritten journal about his overland journey from St. Joseph, Missouri, to Cold Springs, California, in 1850. Occasionally poetic, Cobbey describes buffalo, Indian villages he visited, and bouts of mountain fever in the company. He provides excellent descriptions of the geography and nature of the trail. Cobbey had limited success mining and gives a brief description of conditions in California.

DELANO, ALONZO quotes come from California correspondence from the Ottawa (Illinois) *Free Trader* and the New Orleans *True Delta*, 1849-1952.

HART, WILLIAM HENRY, 1829-1888. Hart's diaries were written between 1852-1888. He made an overland journey by ox team from Quincy, Illinois, to the Humboldt Valley, Nevada, in 1852. He writes about cooking and commerce on the plains, encounters with Indians, and dissatisfaction in his company. Hart mentions a handful of deaths from cholera and includes a moving description of one funeral. Hart also went to California where he mined for gold and worked as a shingle maker for four years. Hart was 23 years old in 1852 when he wrote his memoirs of his travel and life in the area.

NORDHOFF, CHARLES (1830-1901) and his family came to America from Prussia when he was a boy and settled in Cincinnati, Ohio. Winning a reputation as a journalist and writer on the sea, Nordhoff was managing editor of the New York *Evening Post*, 1861-1871. He spent 1872-1873 travelling to California and Hawaii, and returned east to become the Washington correspondent of the New York Herald. He continued to visit California frequently and spent his last years in Coronado. *California: for Health, Pleasure and Residence* (1873) was an extremely popular guidebook that persuaded many to settle in California. It opens with descriptions of the various routes available to the traveller to California and the visitor to Yosemite. Next come suggested points of interest, California agriculture (with hints to prospective settlers), and notes on the Southern California climate.

PLAYER-FROWD, J.G. was an English visitor to California in the early 1870s. *Six Months in California* (1872) is a traveler's guide based on that visit, recounting stays in Omaha, Salt Lake City, the Sierras, Lake Tahoe, Sacramento, San Francisco, Calistoga, Stockton, and the Yosemite Valley. Player-Frowd discusses topics such as California climate, agriculture, mining, and lumbering.

RAE, WILLIAM FRASER, author of *Westward By Rail: The New Route To The East*, D. Appleton & Co., New York: 1871 (Trip made in 1869).

ROOP, ISAAC NEWTON, was born in Carroll County, Maryland, in 1822. Roop started for California in September, 1850. He arrived in San Francisco on the eighteenth of October. He went to Shasta the following June. Stripped of everything but an unconquerable will and being of an adventurous disposition, he turned his back upon civilized life, and journeying across the Sierras, took up his abode in Honey Lake Valley. Honey Lake Valley, as lately as the year 1858, was considered part of Utah Territory. They resolved, in the year 1859, to cut loose from all political communication with Utah. A convention was called, a Constitution drafted, and a new territory was formed and christened Nevada. Isaac N. Roop was chosen Provisional Governor by nearly a unanimous vote.

SHAW, DAVID AUGUSTUS, left Marengo, Illinois, in 1850 for the overland trail to California, where he settled in Pasadena and was an active member of the local Society of Pioneers. *Eldorado* (1900) records Shaw's first stay in the West, 1850-1852, when he worked as a miner and rancher; his return to Illinois and second overland journey west, 1853, this time bringing a herd of horses; and a third round trip to the East, 1856, this time crossing at Panama. In California, Shaw worked as a miner and rancher.

SINCLAIR, JOHN was one of the members of the First Relief [for the Donner Party], who copied the diary of the First Relief, and wrote an account of the Forlorn Hope party while they were in the area of Crystal Lake, December 20, 1846.

SPOONER, ELIJAH ALLEN, 1811-1879 was a Massachusetts native who settled in Kansas in 1857. Spooner farmed and served as a probate judge and county clerk, also prospector who participated in the California Gold Rush of 1849. Most of his letters are addressed to Spooner's wife and were written while on the overland journey by ox team from Adrian, Michigan, to Sacramento, California, in 1849. Spooner writes of encounters with Indians, buffalo hunting, distaste for Sunday travel, a handful of deaths within his company, and traveling conditions. Spooner also mentions his religious faith in most of the letters. Many of his letters describe life in California and his distress that he has not received any letters from home

SQUIER, MIRIAM, (1836-1914), an actress turned journalist who eventually became a powerful figure in American publishing, married publisher Frank Leslie in 1874. In 1877, the couple traveled to California, and Mrs. Leslie recorded details of their luxurious transcontinental rail trip. *California: a pleasure trip from Gotham to the Golden Gate* (1877) chronicles the scenes they passed en route, as well as San Francisco's welcome for the visiting Eastern notables. Her account gives special attention to that city's Chinatown as well as the attractions of Los Angeles and Yosemite. On the return journey, Leslie pictures the desolation of the mining town of Virginia City, Nevada, and the prosperity and progress of Salt Lake City, where she interviewed Brigham Young. The editor's introduction provides details of the attacks brought by publication of the book, with critics exposing Mrs. Leslie's illegitimate birth and complicated marital career.

STEPHENS, LORENZO DOW (b. 1827) was born in New Jersey and raised in Illinois, where he joined a party for California in 1849. *Life sketches of a Jayhawker* (1916) begins with Stephens's overland journey west, including Brigham Young's sermons at the Tabernacle in Salt Lake. He describes prospecting on the Merced River, farming in the Santa Clara Valley, and cattle drives from San Bernadino and San Diego. His memoirs continue through the 1860s, including his part in the 1862 British Columbia gold rush.

THORNTON, JESSE "J" QUINN, wrote *"Oregon and California in 1848,"* which included detailed accounts of the Donner Party, mainly supplied by William Eddy, a member of the party. It was one of the original books dealing with the Donner Party and has since been found to exaggerate the conditions and stories. Whether those exaggerations were Thornton's or Eddy's creations is unknown. William Eddy, 28, was a carriage maker from Belleville, Illinois. With him was his wife Eleanor, 25, and their two children James, 3, and Margaret, 5. The Eddys had one wagon. For more information on the Donner Party, visit: www.DonnerPartyDiary.com

TWAIN, MARK, See Samuel Clemens

VERNEY, EDMUND HOPE, 1838-1910, was a prominent British citizen. Verney served in the Royal Navy for twenty-six years and was elected to four years' service in Parliament. His journal was titled *"An overland journey from San Francisco to New York by way of the Salt Lake City, 1866."*

WILSON, ELIJAH NICHOLAS, (1845-1916) "Uncle Nick" Wilson was taken to Utah by his parents in 1850. At the age of nine he was adopted by Chief Washakie's mother, with whom he lived among the Shoshones for about two years. He then returned to his own family. In 1860, when he was fifteen, he was hired as a Pony Express rider and put in Egan's Division between Shell Creek and Deep Creek. When his Pony Express days were over he avoided towns and thickly populated areas to live on the frontier. His story is available in the book *The White Indian Boy: The Story of Uncle Nick Among the Shoshones* (paperback), by Fredonia Books.

Index

More "Golden" Books

The Golden Corridor is an overview of the life and times in Northern California in the 19th century. It is distinctive because it brings together the unique observations and viewpoints of dozens of people who were in a particular area at a particular time-frame in history.

The Golden Corridor has paved the way for a more in-depth look at various Northern California and Nevada communities, including this one. We made some exciting "finds" in creating these first two books, and there is more to discover, preserve and share. Through today's technology, information that could only be obtained looking through huge historical texts in vast research libraries and archives is now accessible to everyone. And we'll bring it to you.

Does your family have 19th century photos, journals, letters or other documents you'd like to share and preserve?

Let us help you archive your originals electronically. You will receive a copy of all your materials on CD, and you decide if you'd like to retain the originals or entrust them to your local historical society, the Library of Congress, or some other institution for safekeeping.

Please share your family's history. It'll provide one more piece to the complex puzzle that represents our rich culture and background.

Contact us today for details.

More "Golden" books are in the works. They will be even more informative and more fun for readers. They will include more photographic treasures and information from the archives of dozens of smaller historical societies and private collections. And, we're working with the community of historians, historical parks and friends, to discover even more new sources of information.

Over the next couple of years you'll be able to visit many places and see many scenes that few have experienced. Here are some titles and estimated release dates:

The Golden Highway 49	Winter 2006
The Golden Gate - San Francisco	Spring 2007
The Golden Hub - Sacramento	Winter 2007

To pre-order copies of any of these books, or for bulk orders for schools or other organizations, please contact us at:

19thCentury Books / Electric Canvas
1001 Art Road
Pilot Hill, CA 95664
916.933.4490

If you'd like to receive an e-mail announcement when new titles are released, please contact: Jody@19thCentury.us

Thank you for your interest in our rich history. We hope you've enjoyed *The Golden Quest* and will enjoy more of our books in the future.